U0163147

全国专门用途英语(ESP)规划系列教材

冶金科技英语阅读

English Readings of Metallurgical Science and Technology

主　编　康有金

副主编　崔家军　鲁　霞　徐　唱　王妙妙
　　　　吴梦华

编　者　康有金　崔家军　鲁　霞　徐　唱
　　　　王妙妙　刘锦莹　吴　璇　田晶晶
　　　　杨紫琴　刘佳彦　凌　静　吴梦华
　　　　章峰华　李虹仪　贾喻圆　许嘉慧
　　　　卓小艳　潘传惠子

苏州大学出版社

图书在版编目(CIP)数据

冶金科技英语阅读 / 康有金主编. —苏州：苏州大学出版社，2022. 1
全国专门用途英语(ESP)规划系列教材
ISBN 978-7-5672-3859-6

Ⅰ. ①冶… Ⅱ. ①康… Ⅲ. ①冶金工业-英语-阅读教学-教材 Ⅳ. ①TF

中国版本图书馆 CIP 数据核字(2022)第 006411 号

Yejin Keji Yingyu Yuedu
冶金科技英语阅读
书　　名： English Readings of Metallurgical Science and Technology

主　　编： 康有金
责任编辑： 金莉莉
策划编辑： 汤定军
封面设计： 刘　俊

出版发行： 苏州大学出版社(Soochow University Press)
社　　址： 苏州市十梓街 1 号　**邮编：** 215006
印　　装： 广东虎彩云印刷有限公司
网　　址： www. sudapress. com
邮　　箱： sdcbs@ suda. edu. cn
邮购热线： 0512-67480030
销售热线： 0512-67481020

开　　本： 787 mm×1 092 mm　1/16　**印张：** 11　**字数：** 241 千
版　　次： 2022 年 1 月第 1 版
印　　次： 2022 年 1 月第 1 次印刷
书　　号： ISBN 978-7-5672-3859-6
定　　价： 48.00 元

目录 Contents

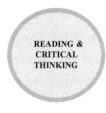

READING & CRITICAL THINKING

LECTURE ONE
Metallurgy

Passage A Metallurgy

Metallurgy is a domain of material science and engineering that studies the physical and chemical behavior of metallic elements, their intermetallic compounds, and their mixtures, which are called alloys. Metallurgy encompasses both the science and the technology of metals. That is how science is applied to the production of metals and the engineering of metal components used in products for both consumers and manufacturers. Metallurgy is distinct from the craft of metalworking. Metalworking relies on metallurgy in a similar manner to how medicine relies on medical science for technical advancement. A specialist practitioner of metallurgy is known as a metallurgist.

The science of metallurgy is subdivided into two broad categories: chemical metallurgy and physical metallurgy. Chemical metallurgy is chiefly concerned with the reduction and oxidation of metals and the chemical performance of metals. Subjects of study in chemical metallurgy include mineral processing, the extraction of metals, thermodynamics, electrochemistry, and chemical degradation (corrosion). In contrast, physical metallurgy focuses on the mechanical properties of metals, the physical properties of metals, and the physical performance of metals. Topics studied in physical metallurgy include crystallography, material characterization, mechanical metallurgy, phase transformations, and failure mechanisms.

Historically, metallurgy has predominately focused on the production of metals. Metal production begins with the processing of ores to extract the metal and includes the mixture of metals to make alloys. Metal alloys are often a blend of at least two different metallic elements. However, nonmetallic elements are often added to alloys in order to achieve properties suitable for an application. The study of metal production is subdivided into ferrous metallurgy (also known as black metallurgy) and nonferrous metallurgy (also known as colored metallurgy). Ferrous metallurgy involves processes and alloys based on iron while nonferrous metallurgy involves processes and alloys based on other metals. The production of ferrous metals accounts for 95 percent of worldwide metal production. Modern metallurgists work in both emerging and traditional areas as part of an interdisciplinary team alongside material scientists and other engineers. Some traditional areas include mineral processing, metal production, heat treatment, failure analysis, and the joining of metals (including welding, brazing, and soldering). Emerging areas for metallurgists include nanotechnology, superconductors, composites, biomedical materials, electronic materials (semiconductors), and surface engineering.

Extraction

Extractive metallurgy is the practice of removing valuable metals from an ore and refining the extracted raw metals into a purer form. To convert a metal oxide or sulphide to a purer metal, the ore must be reduced physically, chemically, or electrolytically.

Extractive metallurgists are interested in three primary streams: feed, concentrate (valuable metal oxide/sulphide) and tailings (waste). After mining, large pieces of the ore feed are broken through crushing or grinding to obtain particles small enough where each particle is either mostly valuable or mostly waste. Concentrating the particles of value in a form supporting separation enables the desired metal to be removed from waste products.

If the ore body and physical environment are conducive to leaching, mining may not be necessary. Leaching dissolves minerals in an ore body and results in an enriched solution. The solution is collected and processed to extract valuable metals.

Ore bodies often contain more than one valuable metal. Tailings of a previous process may be used as a feed in another process to extract a secondary product from the original ore. Additionally, a concentrate may contain more than one valuable metal. That concentrate would then be processed to separate the valuable metals into individual constituents.

Common engineering metals include aluminium, chromium, copper, iron, magnesium, nickel, titanium, zinc, and silicon. These metals are most often used as alloys with the noted exception of silicon. Much effort has been placed on understanding the iron-carbon alloy system, which includes steels and cast irons.

Plain carbon steels (those that contain essentially only carbon as an alloying element) are used in low-cost, high-strength applications where neither weight nor corrosion is a major concern. Cast irons, including ductile irons, are also part of the iron-carbon system.

Stainless steel, particularly austenitic stainless steel, galvanized steel, nickel alloys, titanium alloys, or occasionally copper alloys are used where corrosion resistance is important. Aluminium alloys and magnesium alloys are commonly used when a lightweight strong part is required such as in automotive and aerospace applications.

Copper-nickel alloys (such as Monel) are used in highly corrosive environments and for nonmagnetic applications. Iron-manganese-chromium alloys (Hadfield-type steels) are also used in nonmagnetic applications such as directional drilling. Nickel-based superalloys like Inconel are used in high-temperature applications such as gas turbines, turbochargers, pressure vessels, and heat exchangers. For extremely high temperatures, single crystal alloys are used to minimize creep. In modern electronics, high-purity single crystal silicon is essential for metal-oxide-silicon (MOS) transistors and integrated circuits.

Production

In production engineering, metallurgy is concerned with the production of metallic components for use in consumer or engineering products. This involves the production of alloys, the shaping, the heat treatment, and the surface treatment of the product. Determining the hardness of the metal using the Rockwell, Vickers, and Brinell hardness scales is a commonly used practice that helps better understand the metal's elasticity and plasticity for different applications and production processes. The task of the metallurgist is to achieve a balance between material properties such as cost, weight, strength, toughness, hardness, corrosion, fatigue resistance, and performance in temperature extremes. To achieve this goal, the operating environment must be carefully considered. In a saltwater environment, most ferrous metals and some non-ferrous alloys corrode quickly. Metals exposed to cold or cryogenic conditions may undergo a ductility to brittle transition and lose their toughness, becoming more brittle and prone to cracking. Metals under continual cyclic loading can suffer from metal fatigue. Metals under constant stress at elevated temperatures can creep.

Metalworking processes

Metals are shaped by processes such as:

Casting—molten metal is poured into a shaped mold.

Forging—a red-hot billet is hammered into shape.

Rolling—a billet is passed through successively narrower rollers to create a sheet.

Laser cladding—metallic powder is blown through a movable laser beam (e. g. mounted on an NC 5-axis machine). The resulting melted metal reaches a substrate to form a melt pool. By moving the laser head, it is possible to stack the tracks and build up a three-dimensional piece.

Extrusion—a hot and malleable metal is forced under pressure through a die, which shapes it before it cools.

Sintering—a powdered metal is heated in a non-oxidizing environment after being compressed into a die.

Machining—lathes, milling machines, and drills cut the cold metal to shape.

Fabrication—sheets of metal are cut with guillotines or gas cutters and bent and welded into structural shape.

3D printing—sintering or melting amorphous powder metal in a 3D space to make any object shape.

Cold-working processes, in which the product's shape is altered by rolling, fabrication, or other processes while the product is cold, can increase the strength of the product by a process called work hardening. Work hardening creates microscopic defects in the metal, which resist further changes of shape.

Various forms of casting exist in industry and academia. These include sand casting, investment casting (also called the lost wax process), die casting, and continuous castings. Each of these forms has advantages for certain metals and applications considering factors like magnetism and corrosion.

Heat treatment

Metals can be heat-treated to alter the properties of strength, ductility, toughness, hardness, and corrosion resistance. Common heat treatment processes include annealing, precipitation strengthening, quenching, and tempering. The annealing process softens the metal by heating it and then allowing it to cool very slowly, which gets rid of stresses in the metal and makes the grain structure large and soft-edged so that when the metal is hit or stressed it dents or perhaps bends, rather than breaking; it is also easier to sand, grind, or cut annealed metal. Quenching is the process of cooling high-carbon steel very quickly after heating, thus "freezing" the steel's molecules in the very hard martensite form, which makes the metal harder. There is a balance between hardness and toughness in any steel; the harder the steel, the less tough or impact-resistant it is, and the more impact-resistant it is, the less hard it is. Tempering relieves stresses in the metal that were caused by the hardening process; tempering makes the metal less hard while making it sustain impacts without breaking better.

Often, mechanical and thermal treatments are combined in what is known as thermo-mechanical treatments for better properties and more efficient processing of materials. These processes are common to high-alloy special steels, superalloys, and titanium alloys.

Plating

Electroplating is a chemical surface-treatment technique. It involves bonding a thin layer of another metal such as gold, silver, chromium, or zinc to the surface of the product. This is done by selecting the coating material electrolyte solution which is the material that is going to coat the workpiece (gold, silver, zinc). There needs to be two electrodes of different materials: one the same material as the coating material and one that is receiving the coating material. Two electrodes are electrically charged and the coating material is stuck to the workpiece. It is used to reduce corrosion as well as to improve the product's aesthetic appearance. It is also used to make inexpensive metals look like the more expensive ones (gold, silver).

Shot peening

Shot peening is a cold working process used to finish metal parts. In the process of shot peening, the small round shot is blasted against the surface of the part to be finished. This process is used to prolong the product life of the part, prevent stress corrosion failures, and prevent fatigue. The shot leaves small dimples on the surface like a peen hammer does, which causes compression stress under the dimple. As the shot media strikes the material over and over, it forms many overlapping dimples throughout the piece being treated. The compression stress in the surface of the material strengthens the part and makes it more resistant to fatigue failure, stress failures, corrosion failure, and cracking.

Thermal spraying

Thermal spraying technique is another popular finishing option, and often have high-temperature properties than electroplated coatings. Thermal spraying, also known as a spray welding process, is an industrial coating process that consists of a heat source (flame or other) and a coating material that can be in a powder or wire form which is melted and then sprayed on the surface of the material being treated at a high velocity. The spray treating process is known by many different names such as HVOF (High-Velocity Oxygen Fuel), plasma spray, flame spray, arc spray, and metalizing.

Metallography allows the metallurgist to study the microstructure of metals.

Characterization

Metallurgists study the microscopic and macroscopic structure of metals using metallography, a technique invented by Henry Clifton Sorby. In metallography, an alloy of interest is ground flat and polished to a mirror finish. The sample can then be etched to reveal the microstructure and macrostructure of metals. The sample is then examined in an optical or electron microscope, and the image contrast provides details on the composition, mechanical properties, and processing history.

Crystallography, often using diffraction of X-rays or electrons, is another valuable tool available to the modern metallurgist. Crystallography allows the identification of unknown materials and reveals the crystal structure of the sample. Quantitative crystallography can be used to calculate the number of phases present as well as the degree of strain to which a sample has been subjected.

Terms

1. alloy 合金
2. ferrous metallurgy 黑色冶金
3. nonferrous metallurgy 有色冶金
4. weld 焊接
5. braze 铜焊
6. solder 焊接
7. extractive metallurgy 萃取冶金
8. feed 供给机器的原料;进料
9. concentrate 精矿
10. tailing 尾矿
11. crushing 破碎
12. grinding 研磨
13. ore body 矿体
14. iron-carbon alloy system 铁碳合金系统
15. plain carbon steel 普碳钢
16. ductile iron 球墨铸铁
17. stainless steel 不锈钢
18. galvanized steel 镀锌钢
19. nickel alloy 镍合金
20. titanium alloy 钛合金
21. copper alloy 铜合金
22. aluminium alloy 铝合金
23. magnesium alloy 镁合金
24. copper-nickel alloy 铜镍合金
25. iron-manganese-chromium alloy 钛锰铬合金
26. nickel-based superalloy 镍基高温合金
27. single crystal alloy 单晶合金
28. single crystal silicon 单晶硅
29. metal fatigue 金属疲劳
30. casting 铸造
31. forging 锻造

32. rolling 轧制	33. laser cladding 激光熔覆
34. extrusion 挤压	35. sintering 烧结
36. investment casting 熔模铸造	37. die casting 压铸
38. continuous casting 连续铸造	39. annealing 退火
40. precipitation strengthening 沉淀强化	41. quenching 淬火
42. tempering 回火	43. superalloy 高温合金
44. electroplating 电镀	45. electrolyte solution 电解质溶液
46. shot peening 喷丸处理	47. thermal spraying technique 热喷涂技术
48. electroplated coating 电镀层	49. plasma spray 等离子喷涂
50. flame spray 火焰喷涂	51. arc spray 电弧喷涂

Exercises

1. What's the difference between chemical metallurgy and physical metallurgy?
2. Please describe the process of producing feed, concentrate and tailing.
3. Please compare the advantages of different casting forms.

Passage B An Introduction to Metallurgy

Metallurgy relates to the science and technology of metals. In nature, only a few metals occur in their native state, while all other metals occur in a combined state as their oxides, sulphide, silicates, etc ... The extraction of pure metals from their natural sources is linked to the history of human civilization. Ancient people used the available materials in their environment which includes fire and metals, and they were limited to the metals available on the Earth's surface. In the modern world, we use a wide range of metals in our daily life, which is the result of the development of metallurgical knowledge over thousands of years. Our need for materials with specific properties has led to the production of metallurgy generally dealt with by material scientists and material engineers, who study the physical and chemical behavior of metallic elements, their intermetallic compounds, and their mixtures, which are referred to as alloys. Metallurgy can also be described as the technology of metals, the way in which science is applied to the production of metals and the engineering of metal components for use in products for manufacturers and consumers. The production of metals involves the processing of ores to extract the metal they contain, and the mixing of metals or other elements to produce many metal alloys. It is essential to design an eco-friendly metallurgical process that would minimize waste, maximize energy efficiency. Such advances in metallurgy are vital for the economic and technical progress in the current era.

History of metallurgy

The present-day use of metals is the culmination of a long path of development extending over approximately 6,500 years. It is generally agreed that the first known metals were gold, silver, and copper, which occurred in the native or metallic state, of which the earliest were in all probability nuggets of gold found in the sands and gravels of riverbeds. Such native metals became known and were appreciated for their ornamental and utilitarian values during the latter part of the Stone Age.

Earliest development

Gold can be agglomerated into larger pieces by cold hammering, but native copper cannot, and an essential step toward the Metal Age was the discovery that metals such as copper could be fashioned into shapes by melting and casting in molds; among the earliest known products of this type are copper axes cast in the Balkans in the 4th millennium BCE. Another step was the discovery that metals could be recovered from metal-bearing minerals. These had been collected and could be distinguished on the basis of colour, texture, weight, and flame colour and smell when heated. The notably greater yield obtained by heating native copper with associated oxide minerals may have led to the smelting process since these oxides are easily reduced to metal in a charcoal bed at temperatures over 700 ℃ (1,300 ℉), as the reductant, carbon monoxide, becomes increasingly stable. To affect the agglomeration and separation of melted or smelted copper from its associated minerals, it was necessary to introduce iron oxide as a flux. This further step forward can be attributed to the presence of iron oxide gossan minerals in the weathered upper zones of copper sulfide deposits.

Contemporary metallurgy from 1850

The technical advances of the second industrial revolution would make the hegemony of iron metallurgy even greater. Where in the 18th century the British had allowed for the industrialization of the cast and wrought iron production, in the second half of the 19th century, decisive technical changes would be achieved in the steel industry. By means of the Bessemer (1856), Siemens-Martin (1857 - 1864), and Thomas-Gilchrist (1878) converters, it was possible to produce steel in large quantities at prices far below the previous ones. Thus steel could be used, in place of wrought iron and other materials, in a much more extensive way in all types of applications. Thanks to its hardness and elasticity, it could be incorporated into the building, railroad, shipbuilding, and mechanical industries, among other sectors, contributing in a decisive way to the advancement of industrial societies.

The consumption of nonferrous metals was also affected by advances in the industry. New products and new demand sectors—such as those linked to the modern armaments industry, electrification, the beginnings of the automotive and the aeronautics sectors, and certain goods produced for domestic consumption—increased the general demand for metals and alloys with more and more precise technical requirements. The new colonialism, the internationalization of the economy, and the development of modern means of transport led to the discovery of new ore deposits and the extension of mining throughout the world. In certain cases, the procedures used in the iron and steel industry could be applied to other metals sucessfully. But limitations imposed by traditional metallurgy were overcome with the development of new methods of electrometallurgy. These consisted of applying the potential of electric power as a thermal agent and incorporating electrolytic methods in the extraction and refined processes. The new procedures made it possible to improve the purity of the metals obtained and to produce new metals and alloys industrially.

Although there had already been advances in the industrial chemistry at the beginning of the 19th century, it was impossible to develop the full potential of electrometallurgy until the availability of abundant and cheap electricity. In fact, copper, whose purity was a decisive characteristic for use as an electrical conductor, was the first industrially refined metal obtained by means of electrolysis (1865), which contributed in a decisive way to the growth of the electric industry. Aluminum, magnesium, special steels, and other very diverse alloys could be developed commercially thanks to the technical and economical possibilities of electrometallurgy. In addition, electric furnaces have provided the industry with greater flexibility of location and have reduced pollution emissions considerably.

The development of contemporary metallurgy has been associated with important advances in the administration and organization of the industry. The necessity of integrating different production processes in a single factory, the high cost of equipment, and the concentration of production and markets have transformed the structure of the industry and paved the way for the great modern firms and mass production, especially in the iron and steel industry. Although in some cases experimentation and practice continue to be important, the technological development in contemporary metallurgy depends more and more on advances in science and investigation.

During the 20th century, the extraction and treatment of metals continued to evolve without interruption, and at an increasingly rapid pace. The progress of contemporary metallurgy has been based on new industries and demand sectors, such as atomic engineering, electronics, telecommunications, and the aerospace and military industries. Improvement in the quality of the final products, cost reductions, the search for new alloys, and industrial production of new metals like titanium, beryllium, zirconium, and silicon have required the introduction of constant technical changes, as much in the production and finishing processes as in methods of analysis and control of the production process. Some of the technological alternatives developed in the new metallurgy of the twentieth century are continuous casting and the Basic Oxygen Steel process in the steelmaking industry, hydrometallurgy (aqueous processing of metals) for the treatment of precious metals and minerals of high value (uranium, nickel, cobalt), the use of powder metallurgy to obtain certain structural parts for industrial use, and special alloys of the mixed constitution (with metallic and nonmetallic elements).

The complexity and speed of technological change in advanced societies make it impossible to foresee future tendencies in metallurgy. The most recent progress in the science and engineering of materials is allowing the constant development of new materials and processes. Although in some cases new plastics, ceramics, and hybrid materials may compete with and displace metals in certain applications, the development of metallurgical engineering in the current technological environment of the cross-disciplinary investigation will continue to constitute a decisive element in the technological progress and industrial systems of the future.

Metallurgy in production engineering

In production engineering, metallurgy is concerned with the production of metallic components for use in consumer or engineering products. This involves the production of alloys, the shaping, the heat treatment, and the surface treatment of the product. The task of the metallurgist is to achieve design criteria specified by the mechanical engineer, such as cost, weight, strength, toughness, hardness, corrosion and fatigue resistance, and performance in temperature extremes.

Common engineering metals are aluminum, chromium, copper, iron, magnesium, nickel, titanium, and zinc. These are most often used as alloys. Much effort has been placed on understanding a very important alloy system, that of purified iron, which has carbon dissolved in it, better known as steel. Normal steel is used in low-cost, high-strength applications where weight and corrosion are not a problem. Cast irons, including ductile irons are also part of this system.

Stainless steel or galvanized steel is used where corrosion resistance is important. Aluminium alloys and magnesium alloys are used for applications where strength and lightness are required. Most engineering metals are stronger than most plastics and tougher than most ceramics. Composites of plastics and materials such as glass fiber and carbon fiber rival metals in applications requiring high tensile strength with little weight. Concrete rivals metals in applications requiring high compressive strength and resistance to the effects of water. Wood rivals metal in applications requiring low cost and availability of materials and low cost of construction, as well as in applications requiring certain aesthetics.

The operating environment of the product is very important—a well-designed material will resist expected failure modes such as corrosion, stress concentration, metal fatigue, creep, and environmental stress fracture. Ferrous metals and some aluminium alloys in water and especially in an electrolytic solution such as seawater, corrode quickly. Metals in cold or cryogenic conditions tend to lose their toughness, becoming more brittle and prone to cracking. Metals under continual cyclic loading can suffer from metal fatigue. Metals under constant stress in hot conditions can creep.

Metallurgical techniques

Metallurgists study the microscopic and macroscopic mechanisms that cause the metal or alloy to behave in the way that it does—that is, the changes that occur on the atomic level that affect the metal's (or alloy's) macroscopic properties. Examples of tools used for microscopic examination of metals are optical.

Metallurgists study the crystallography, the effects of temperature and heat treatment on the component phases of alloys, such as the eutectic and the properties of those alloy phases. The macroscopic properties of metals are tested using machines and devices that measure tensile strength, compressive strength, and hardness.

New discovery in metallurgy

When and where did humans invent metal smelting? Scientists from Heidelberg University, London, and Cambridge (Great Britain) have found the answer to this long-debated question in the history of technology. Metallurgy does not have a single origin but probably arose at various locations at about the same time. The experts reached this conclusion after reexamining the 8,500-year-old copper slag and analysing the chemical composition of other copper artefacts from the Stone Age settlement of Çatalhöyük in the Near East.

This history of civilization is broadly divided into the Stone, Bronze, and Iron Ages. What is little known, however, is that the copper metal was already being processed at the beginning of the Neolithic Age approximately 10,000 years ago in the Fertile Crescent from the Levant through East Anatolia to the Zagros Mountains in Iran. "But because this is all pure, native copper, we can't really call it true metallurgy," explains Prof. Dr. Ernst Pernicka, Scientific Director of Heidelberg University's Curt Engelhorn Centre for Archaeometry, which is located in Mannheim. Copper was found in nature as metal, and according to the researchers, it was probably considered a special type of stone. Because the production of metal from ore ushers in a new era in human history, it is important to know when and where it first developed and whether metal smelting really originated in a single location.

For a long time, a small amount of copper slag from the Neolithic site at Çatalhöyük was thought to be the earliest evidence for the pyrometallurgical extraction of copper from ore. This settlement existed from approximately 7,100 to 6,000 BC and is considered to be the most significant site in the Near East that affords any insight into the development of human habitats. The copper slag was located in layers that dated back to 6,500 years before Christ and was therefore 1,500 years older than the world's earliest evidence of copper smelting. "The find seemed to point to the birthplace of metallurgy, with the technology slowly spreading from there in all directions," states Prof. Pernicka, who also heads the Archaeometry and Archaeometallurgy Research Group at the Institute of Earth Sciences at Heidelberg University.

However new studies, including those at the Curt Engelhorn Centre for Archaeometry, have demonstrated that the slag was an unintentional by-product of a domestic fire. The extreme heat of the fire slagged the green copper ores, which were used as the pigment. The slag differs in chemical composition from another artefact from the Stone Age settlement, a bead of folded sheets made of pure copper. This threw new light on the scientific significance of the copper slag of Çatalhöyük, now putting the earliest known examples of the copper extraction from ore around 5,000 BC in Southeast Europe and Iran. The current findings indicate that this revolutionary development in humankind probably came about at roughly the same time but in multiple locations. "We thereby solved a controversial problem in the history of technology," underscores Prof. Pernicka.

Terms

1. agglomeration 团聚
2. flux 助熔剂
3. nonferrous metal 有色金属
4. electrometallurgy 电冶金
5. titanium 钛
6. beryllium 铍
7. zirconium 锆
8. uranium 铀
9. nickel 镍
10. cobalt 钴
11. crystallography 晶体学
12. copper slag 铜渣

Exercises

1. Please briefly summarize the development of metallurgy after the Second Industrial Revolution.

2. Please describe characteristics and application scopes of different steels and alloys according to the passage and network information.

3. When and in which countries did metallurgy originate?

READING & CRITICAL THINKING

LECTURE TWO
Metals

1. Iron

Passage A An Overview of Iron

Iron is a chemical element with the symbol Fe (from Latin *ferrum*) and atomic number 26. It is a metal that belongs to the first transition series and group 8 of the periodic table. It is by mass the most common element on Earth, forming much of Earth's outer and inner core. It is the fourth most common element in the Earth's crust.

In its metallic state, iron is rare in the Earth's crust, limited to deposition by meteorites. Iron ores, by contrast, are among the most abundant in the Earth's crust, although extracting usable metal from them requires kilns or furnaces capable of reaching 1,500 ℃ (2,732 ℉) or higher, about 500 ℃ (932 ℉) higher than what is enough to smelt copper. Humans started to master that process in Eurasia only about 2,000 BCE, and the use of iron tools and weapons began to displace copper alloys, in some regions, only around 1,200 BCE. That event is considered the transition from the Bronze Age to the Iron Age. In the modern world, iron alloys, such as steel, cast iron, and special steels are by far the most common industrial metals, because of their high mechanical properties and low cost.

Pristine and smooth pure iron surfaces are mirror-like silvery-gray. However, iron reacts readily with oxygen and water to produce black hydrated iron oxides, commonly known as rust. Unlike the oxides of some other metals, which form passivating layers, rust occupies more volume than the metal and thus flakes off, exposing fresh surfaces for corrosion.

The body of an adult human contains about 4 grams (0.005% body weight) of iron, mostly in hemoglobin and myoglobin. These two proteins play essential roles in vertebrate metabolism, respectively oxygen transport by blood and oxygen storage in muscles. To maintain the necessary levels, human iron metabolism requires a minimum of iron in the diet. Iron is also the metal at the active site of many important redox enzymes dealing with cellular respiration and oxidation and reduction in plants and animals.

Chemically, the most common oxidation states of iron are iron (II) and iron (III). Iron shares many properties of other transition metals, including the other group 8 elements, ruthenium, and osmium. Iron forms compounds in a wide range of oxidation states, −2 to +7. Iron also forms many coordination compounds; some of them, such as ferrocene, ferrioxalate, and Prussian blue, have substantial industrial, medical, or research applications.

Terms

1. crust 地壳	2. kiln 窑，炉
3. furnace 熔炉	4. copper 铜
5. Eurasia 亚欧大陆	6. Bronze Age 青铜器时代
7. Iron Age 铁器时代	8. rust 锈
9. passivate 使钝化	10. hemoglobin 血红蛋白
11. myoglobin 肌红蛋白	12. protein 蛋白质
13. vertebrate 脊椎动物	14. metabolism 新陈代谢
15. redox enzyme 氧化还原酶	16. cellular respiration 细胞呼吸
17. reduction 还原	18. ruthenium 钌
19. osmium 锇	20. ferrocene 二茂铁
21. ferrioxalate 草酸铁	22. Prussian blue 普鲁士蓝

Exercises

1. What's the symbol of iron's chemical element?
2. What is iron's appearance?
3. What is iron's melting point?

Passage B Characteristics

Allotropes

At least four allotropes of iron (differing atom arrangements in the solid) are known, conventionally denoted α, γ, δ, and ε.

The first three forms are observed at ordinary pressures. As molten iron cools past its freezing point of 1,538 ℃, it crystallizes into its δ allotrope, which has a body-centered cubic crystal structure. As it cools further to 1,394 ℃, it changes to its γ-iron allotrope, a face-centered cubic crystal structure, or austenite. At 912 ℃ and below, the crystal structure again becomes the bcc α-iron allotrope.

The physical properties of iron at very high pressures and temperatures have also been studied extensively, because of their relevance to theories about the cores of the Earth and other planets. Above approximately 10 GPa and temperatures of a few hundred or less, α-iron changes into another hexagonal close-packed structure, which is also known as ε-iron. The higher-temperature γ-phase also changes into ε-iron but does so at higher pressures.

Some controversial experimental evidence exists for a stable β-phase at pressures above 50 GPa and temperatures of at least 1,500 K. It is supposed to have an orthorhombic or a double hcp structure. Confusingly, the term "β-iron" is sometimes also used to refer to α-iron above its Curie point, when it changes from being ferromagnetic to paramagnetic, even though its crystal structure has not changed.

The inner core of the Earth is generally presumed to consist of an iron-nickel alloy with ε (or β) structure.

Melting and boiling points

The melting and boiling points of iron, along with its enthalpy of atomization, are lower than those of the earlier 3d elements from scandium to chromium, showing the lessened contribution of the 3d electrons to metallic bonding as they are attracted more and more into the inert core by the nucleus; however, they are higher than the values for the previous element manganese because that element has a half-filled 3d subshell and consequently its d-electrons are not easily delocalized. This same trend appears for ruthenium but not osmium.

The melting point of iron is experimentally well defined for pressures less than 50

GPa. For greater pressures, published data (as of 2007) still varies by tens of gigapascals and over a thousand kelvin.

Magnetic properties

Below its Curie point of 770 ℃, α-iron changes from paramagnetic to ferromagnetic: the spins of the two unpaired electrons in each atom generally align with the spins of its neighbors, creating an overall magnetic field. This happens because the orbitals of those two electrons (dz2 and dx2 y2) do not point toward neighboring atoms in the lattice, and therefore are not involved in metallic bonding.

In the absence of an external source of the magnetic field, the atoms get spontaneously partitioned into magnetic domains, about 10 micrometers across, such that the atoms in each domain have parallel spins, but different domains have other orientations. Thus, a macroscopic piece of iron will have a nearly zero overall magnetic field.

The application of an external magnetic field causes the domains that are magnetized in the same general direction to grow at the expense of adjacent ones that point in other directions, reinforcing the external field. This effect is exploited in devices that need to channel magnetic fields, such as electrical transformers, magnetic recording heads, and electric motors. Impurities, lattice defects, or grain and particle boundaries can "pin" the domains in the new positions, so that the effect persists even after the external field is removed—thus turning the iron object into a (permanent) magnet.

Similar behavior is exhibited by some iron compounds, such as the ferrites and the mineral magnetite, a crystalline form of the mixed iron (II, III) oxide Fe_3O_4 (although the atomic-scale mechanism, ferrimagnetism, is somewhat different). Pieces of magnetite with natural permanent magnetization (lodestones) provided the earliest compasses for navigation. Particles of magnetite were extensively used in magnetic recording media such as core memories, magnetic tapes, floppies, and disks until they were replaced by cobalt-based materials.

Isotopes

Iron has four stable isotopes: ^{54}Fe (5.845% of natural iron), ^{56}Fe (91.754%), ^{57}Fe (2.119%), and ^{58}Fe (0.282%). 20 - 30 artificial isotopes have also been created. Of these stable isotopes, only ^{57}Fe has a nuclear spin (-12). The nuclide ^{54}Fe theoretically can undergo double electron capture to ^{54}Cr, but the process has never been observed and only a lower limit on the half-life of 3.1 ×1,022 years has been established.

^{60}Fe is an extinct radionuclide of a long half-life (2.6 million years). It is not found on Earth, but its ultimate decay product is its granddaughter, the stable nuclide ^{60}Ni. Much of the past work the isotopic composition of iron has focused on the nucleosynthesis of ^{60}Fe through studies of meteorites and ore formation. In the last decade, advances in mass spectrometry have allowed the detection and quantification of minute, naturally occurring variations in the ratios of the stable isotopes of iron. Much of this work is driven by the Earth and planetary science communities, although applications to biological and industrial systems are emerging.

In phases of the meteorites Semarkona and Chervony Kut, a correlation between the concentration of ^{60}Ni, the granddaughter of ^{60}Fe, and the abundance of the stable iron isotopes provided evidence for the existence of ^{60}Fe at the time of formation of the Solar System. Possibly, the energy released by the decay of ^{60}Fe, along with that released by ^{26}Al, contributed to the remelting and differentiation of asteroids after their formation 4.6 billion years ago. The abundance of ^{60}Ni present in extraterrestrial material may bring further insight into the origin and early history of the Solar System.

The most abundant iron isotope ^{56}Fe is of particular interest to nuclear scientists because it represents the most common endpoint of nucleosynthesis. Since ^{56}Ni (14 alpha particles) is easily produced from lighter nuclei in the alpha process in nuclear reactions in supernovae (see silicon burning process), it is the endpoint of fusion chains inside extremely massive stars, since the addition of another alpha particle, resulting in ^{60}Zn, requires a great deal of more energy. This ^{56}Ni, which has a half-life of about 6 days, is created in quantity in these stars, but soon decays by two successive positron emissions within supernova decay products in the supernova remnant gas cloud, first to radioactive ^{56}Co, and then to stable ^{56}Fe. As such, iron is the most abundant element in the core of red giants and is the most abundant metal in iron meteorites and the dense metal cores of planets such as Earth. It is also very common in the universe, relative to other stable metals of approximately the same atomic weight. Iron is the sixth most abundant element in the Universe and the most common refractory element.

Although a further tiny energy gain could be extracted by synthesizing ^{62}Ni, which has a marginally higher binding energy than ^{56}Fe, conditions in stars are unsuitable for this process. Element production in supernovas and distribution on Earth greatly favor iron over nickel, and in any case, ^{56}Fe still has a lower mass per nucleon than ^{62}Ni due to its higher fraction of lighter protons. Hence, elements heavier than iron require a supernova for their formation, involving rapid neutron capture by starting ^{56}Fe nuclei.

In the far future of the universe, assuming that proton decay does not occur, cold fusion occurring via quantum tunneling would cause the light nuclei in ordinary matter to fuse into ^{56}Fe nuclei. Fission and alpha-particle emission would then make heavy nuclei decay into iron, converting all stellar-mass objects to cold spheres of pure iron.

Terms

1. allotrope 同素异形体
2. face-centered cubic 面心立方
3. austenite 奥氏体
4. orthorhombic 正交(晶)的,斜方晶系的
5. iron-nickel alloy 铁镍合金
6. enthalpy of atomization 原子化焓
7. scandium 钪
8. chromium 铬
9. electron 电子
10. nucleus 核;原子核;细胞核
11. manganese 锰
12. Curie point 居里点
13. ferromagnetic 铁磁性的
14. magnetic field 磁场
15. magnetic domain 磁畴;磁区
16. electrical transformer 电力变压器
17. magnetic recording head 磁记录头
18. electric motor 电动机
19. lattice defect 晶格缺陷
20. particle 微粒
21. mineral magnetite 矿物磁铁矿
22. core memory 核心记忆
23. magnetic tape 磁带
24. floppy 软磁盘
25. disk 磁盘;磁碟
26. isotope 同位素
27. nuclear spin 核自旋
28. nuclide 核素
29. double electron capture 双电子俘获
30. extinct radionuclide 熄灭的放射性核素
31. half-life (放射性物质的)半衰期
32. Ni 镍
33. body-centered cubic crystal structure 体心立方晶体结构
34. hexagonal close-packed structure 密排六方结构
35. ferrimagnetism 铁氧体磁性
36. compass 指南针;圆规;范围;范畴;界限

Exercises

1. What are the melting and boiling points?
2. How many iron's stable isotopes does iron have and what are they?
3. What is the most abundant iron isotope?

Passage C Origin and Occurrence in Nature

Cosmogenesis

Iron's abundance in rocky planets like Earth is due to its abundant production by fusion in high-mass stars, where it is the last element to be produced with the release of energy before the violent collapse of a supernova, which scatters the iron into space.

Metallic iron

Metallic or native iron is rarely found on the surface of the Earth because it tends to oxidise. However, both the Earth's inner and outer core, which account for 35% of the mass of the whole Earth, are believed to consist largely of an iron alloy, possibly with nickel. Electric currents in the liquid outer core are believed to be the origin of the Earth's magnetic field. The other terrestrial planets (Mercury, Venus, and Mars) as well as the Moon are believed to have a metallic core consisting mostly of iron. The M-type asteroids are also believed to be partly or mostly made of metallic iron alloy.

The rare iron meteorites are the main form of natural metallic iron on the Earth's surface. Items made of cold-worked meteoritic iron have been found in various archaeological sites dating from a time when iron smelting had not yet been developed, and the Inuit in Greenland have been reported to use iron from the Cape York meteorite for tools and hunting weapons. About 1 in 20 meteorites consists of the unique iron-nickel minerals taenite (35% – 80% iron) and kamacite (90% – 95% iron).

Mantle minerals

Ferropericlase (Mg, Fe) O, a solid solution of periclase (MgO) and wustite (FeO), makes up about 20% of the volume of the lower mantle of the Earth, which makes it the second most abundant mineral phase in that region after silicate perovskite (Mg, Fe) SiO_3; it is also the major host for iron in the lower mantle. At the bottom of the transition zone of the mantle, the reaction γ-(Mg, Fe) [SiO_4] θ (Mg, Fe) [SiO_3] + (Mg, Fe) O transforms γ-olivine into a mixture of silicate perovskite and ferropericlase and vice versa. In the literature, this mineral phase of the lower mantle is also often called magnesiowustite. Silicate perovskite may form up to 93% of the lower mantle, and the magnesium iron form, (Mg, Fe) SiO_3, is considered to be the most abundant mineral in the Earth, making up 38% of its volume.

Earth's crust

While iron is the most abundant element on the Earth, it accounts for only 5% of the Earth's crust; thus, being only the fourth most abundant element, after oxygen, silicon, and aluminium.

Most of the iron in the crust is combined with various other elements to form many iron minerals. An important class is the iron oxide minerals such as hematite (Fe_2O_3), magnetite (Fe_3O_4), and siderite ($FeCO_3$), which are the major ores of iron. Many igneous rocks also contain the sulfide minerals pyrrhotite and pentlandite. During weathering, iron tends to leach from sulfide deposits as the sulfate and from silicate deposits as the bicarbonate. Both of these are oxidised in an aqueous solution and precipitate in even mildly elevated pH as iron (III) oxide.

Large deposits of iron are banded iron formations, a type of rock consisting of repeated thin layers of iron oxides alternating with bands of iron-poor shale and chert. The banded iron formations were laid down in the time between 3,700 million years ago and 1,800 million years ago.

Materials containing finely ground iron (III) oxides or oxidehydroxides, such as ochre, have been used as yellow, red, and brown pigments since prehistorical times. They contribute as well to the color of various rocks and clays, including entire geological formations like the Painted Hills in Oregon and the Buntsandstein. Through Eisensandstein (a jurassic "iron sandstone", e.g. from Donzdorf in Germany) and Bath stone in the UK, iron compounds are responsible for the yellowish color of many historical buildings and sculptures. The proverbial red color of the surface of Mars is derived from an iron oxide-rich regolith.

Significant amounts of iron occur in the iron sulfide mineral pyrite (FeS_2), but it is difficult to extract iron from it and it is therefore not exploited. In fact, iron is so common that production generally focuses only on ores with very high quantities of it.

According to the International Resource Panel's Metal Stocks in Society report, the global stock of iron in use in society is 2,200 kg per capita. More-developed countries differ in this aspect from less-developed countries (7,000 - 14,000 vs 2,000 kg per capita).

Terms

1. γ-olivine 伽马橄榄石
2. aluminium 铝
3. iron oxide mineral 铁矿石,氧化铁
4. hematite 赤铁矿
5. magnetite 磁铁矿
6. siderite 菱铁矿;铁陨星,陨铁
7. ore of iron 铁矿石
8. igneous rock 火成岩
9. pyrrhotite 磁黄铁矿
10. pentlandite 镍黄铁矿,硫镍铁矿
11. weathering 风化;侵蚀;雨蚀
12. iron oxide 氧化铁
13. banded iron formation 带状铁构造
14. shale 页岩
15. chert 燧石
16. ochre 赭石;赭色
17. pigment 颜料
18. clay 黏土;泥土;陶土
19. Painted Hills 彩绘山
20. Buntsandstein 斑砂岩统
21. jurassic 侏罗纪的;侏罗纪岩系的
22. Donzdorf 栋茨多夫(德国城市)
23. Bath stone 巴斯石;巴斯石头;巴斯岩
24. regolith 风化层;表皮土
25. pyrite 黄铁矿
26. International Resource Panel's Metal Stocks in Society report 国际资源小组的社会金属库存报告

Exercises

1. Where is iron's abundance?
2. How much does iron account for in the Earth's crust?
3. What is iron's occurrence in nature?

Passage D Chemistry and Compounds

Iron shows the characteristic chemical properties of the transition metals, namely the ability to form variable oxidation states differing by steps of one and a very large coordination and organometallic chemistry: indeed, it was the discovery of an iron compound, ferrocene, which revolutionalized the latter field in the 1950s. Iron is sometimes considered as a prototype for the entire block of transition metals, due to its abundance and the immense role it has played in the technological progress of humanity. Its 26 electrons are arranged in the configuration $[Ar]^3d6^4s^2$, of which the 3d and 4s electrons are relatively close in energy, and thus it can lose a variable number of electrons and there is no clear point where further ionization becomes unprofitable.

Iron forms compound mainly in the oxidation states +2 [iron (II), "ferrous"] and +3 [iron (III), "ferric"]. Iron also occurs in higher oxidation states, e. g. the purple potassium ferrate (K_2FeO_4), which contains iron in its +6 oxidation state. Although iron (VIII) oxide (FeO_4) has been claimed, the report could not be reproduced and such a species (at least with iron in its +8 oxidation state) has been found to be improbable computationally. However, one form of anionic $[FeO_4]$ – with iron in its +7 oxidation state, along with an iron(V) – peroxo isomer, has been detected by infrared spectroscopy at 4 K after cocondensation of laser-ablated Fe atoms with a mixture of O_2/Ar. Iron (IV) is a common intermediate in many biochemical oxidation reactions. Numerous organoiron compounds contain formal oxidation states of +1, 0, −1, or even −2. The oxidation states and other bonding properties are often assessed using the technique of Mössbauer spectroscopy. Many mixed-valence compounds contain both iron (II) and iron (III) centers, such as magnetite and Prussian blue. The latter is used as the traditional "blue" in blueprints.

Iron is the first of the transition metals that cannot reach its group oxidation state of +8, although its heavier congeners ruthenium and osmium can, with ruthenium having more difficulty than osmium. Ruthenium exhibits aqueous cationic chemistry in its low oxidation states similar to that of iron, but osmium does not, favoring high oxidation states in which it forms anionic complexes. In the second half of the 3d transition series, vertical similarities down the groups compete with the horizontal similarities of iron with its neighbor's cobalt and nickel in the periodic table, which are also ferromagnetic at room temperature and share similar chemistry. As such, iron, cobalt, and nickel are sometimes grouped together as the iron triad.

Unlike many other metals, iron does not form amalgams with mercury. As a result, mercury is traded in standardized 76-pound flasks made of iron.

Iron is by far the most reactive element in its group; it is pyrophoric when finely divided and dissolves easily in dilute acids, giving Fe^{2+}. However, it does not react with concentrated nitric acid and other oxidizing acids due to the formation of an impervious oxide layer, which can nevertheless react with hydrochloric acid.

Oxides and hydroxides

Iron forms various oxide and hydroxide compounds; the most common are iron (II, III) oxide (Fe_3O_4), and iron (III) oxide (Fe_2O_3). Iron (II) oxide also exists, though it is unstable at room temperature. Despite their names, they are actually all non-stoichiometric compounds whose compositions may vary. These oxides are the principal ores for the production of iron (see bloomery and blast furnace). They are also used in the production of ferrites, useful magnetic storage media in computers, and pigments. The best-known sulfide is iron pyrite (FeS_2), also known as fool's gold owing to its golden luster. It is not an iron(IV) compound, but is actually an iron(II) polysulfide containing Fe^{2+} and S^{2-} ions in a distorted sodium chloride structure.

Coordination compounds

Due to its electronic structure, iron has very large coordination and organometallic chemistry.

Many coordination compounds of iron are known. A typical six-coordinate anion is hexachloroferrate (III), $[FeCl_6]^{3-}$, found in the mixed salt tetrakis (methylammonium) hexachloroferrate (III) chloride. Complexes with multiple bidentate ligands have geometric isomers. For example, thetrans-chlorohydridobis [bis-1,2-(diphenylphosphino) ethane] iron (II) complex is used as a starting material for compounds with the $Fe(dppe)_2$ moiety. The ferrioxalate ion with three oxalate ligands (shown at right) displays helical chirality with its two non-superposable geometries labeled for the lef-handed screw axis and Δ (delta) for the right-handed screw axis, in line with IUPAC conventions. Potassium-ferrioxalate is used in chemical actinometry and along with its sodium salt undergoes photoreduction applied in old-style photographic processes. The dihydrate of iron (II)-oxalate has a polymeric structure with co-planar oxalate ions bridging between iron with the water of crystallization located forming the caps of each octahedron, as illustrated below.

Iron (III) complexes are quite similar to chromium (III) complexes except for iron(III)'s preference for O-donor instead of N-donor ligands. The latter tend to be rather more unstable than iron (II) complexes and often dissociate in water. Many Fe-O complexes show intense colors and are used as tests for phenols or enols. For example, in the ferric chloride test, used to determine the presence of phenols, iron (III) chloride reacts with a phenol to form a deep violet complex:

$$3ArOH + FeCl_3 \rightarrow Fe(OAr)_3 + 3HCl \ (Ar = aryl)$$

Among the halide and pseudohalide complexes, fluoro complexes of iron (III) are the most stable, with the colorless $[FeF_5(H_2O)]^{2-}$ being the most stable in aqueous solution. Chloro complexes are less stable and favor tetrahedral coordination as in $[FeCl_4]^-$; $[FeBr_4]^-$ and $[FeI_4]^-$ are reduced to iron (II) easily. Thiocyanate is a common test for the presence of iron (III) as it forms the blood-red $[Fe(SCN)(H_2O)_5]^{2+}$. Like manganese (II), most iron (III) complexes are high-spin, the exceptions being those with ligands that are high in the spectrochemical series such as cyanide. An example of a low-spin iron (III) complex is $[Fe(CN)_6]^{3-}$. The cyanide ligands may easily be detached in $[Fe(CN)_6]^{3-}$, and hence this complex is poisonous, unlike the iron (II) complex $[Fe(CN)_6]^{4-}$ found in Prussian blue, which does not release hydrogen cyanide except when dilute acids are added. Iron shows a great variety of electronic spinstates, including every possible spin quantum number value for a d-block element from 0 (diamagnetic) to 5/2 (5 unpaired electrons). This value is always half the number of unpaired electrons. Complexes with zero to two unpaired electrons are considered low-spin and those with four or five are considered high-spin.

Iron (II) complexes are less stable than iron (III) complexes but the preference for O-donor ligands is less marked, so that for example, $[Fe(NH_3)_6]^{2+}$ is known while $[Fe(NH_3)_6]^{3+}$ is not. They have a tendency to be oxidised to iron (III) but this can be moderated by low pH and the specific ligands used.

Organometallic compounds

Organoiron chemistry is the study of organometallic compounds of iron, where carbon atoms are covalently bound to the metal atom. They are many and varied, including cyanide complexes, carbonyl complexes, sandwich and half-sandwich compounds.

Prussian blue or "ferric ferrocyanide", $Fe_4[Fe(CN)_6]_3$, is an old and well-known iron-cyanide complex, extensively used as a pigment and in several other applications. Its formation can be used as a simple wet chemistry test to distinguish between aqueous solutions of Fe^{2+} and Fe^{3+} as they react with potassium ferricyanide and potassium ferrocyanide to form Prussian blue, respectively.

Another old example of an organoiron compound is iron pentacarbonyl, $Fe(CO)_5$, in which a neutral iron atom is bound to the carbon atoms of five carbon monoxide molecules. The compound can be used to make carbonyl iron powder, a highly reactive form of metallic iron. Thermolysis of iron pentacarbonyl gives triirondodecacarbonyl, $Fe_3(CO)_{12}$, with a cluster of three iron atoms at its core. Collman's reagent, disodium tetracarbonylferrate, is a useful reagent for organic chemistry; it contains iron in the −2 oxidation state. Cyclopentadienyliron dicarbonyl dimer contains iron in the rare +1 oxidation state.

A landmark in this field was the discovery in 1951 of the remarkably stable sandwich compound ferrocene $Fe(C_5H_5)_2$, by Paulson and Kealy and independently by Miller and others, whose surprising molecular structure was determined only a year later by Woodward and Wilkinson and Fischer. Ferrocene is still one of the most important tools and models in this class.

Iron-centered organometallic species are used as catalysts. The Knölker complex, for example, is a transfer hydrogenation catalyst for ketones.

Industrial uses

The iron compounds produced on the largest scale in industry are iron (II) sulfate ($FeSO_4 \cdot 7H_2O$) and iron(III) chloride ($FeCl_3$). The former is one of the most readily available sources of iron (II), but is less stable to aerial oxidation than Mohr's salt [$(NH_4)2Fe(SO_4)_2 \cdot 6H_2O$]. Iron (II) compounds tend to be oxidised to iron (III) compounds in the air.

Terms

1. phenol 酚
2. enol 烯醇
3. ferric chloride test 三氯化铁试验
4. thiocyanate 硫氰酸盐(或酯)
5. cyanide 氰化物
6. hydrogen cyanide 氰化氢
7. organoiron chemistry 有机铁化学
8. organometallic compound 金属有机化合物

9. carbonyl complex 羰基络合物
10. potassium ferricyanide 铁氰化钾
11. potassium ferrocyanide 亚铁氰化钾
12. iron pentacarbonyl 五羰铁
13. carbonyl iron 羰基铁
14. triirondodecacarbonyl 十二羰基三铁
15. disodium tetracarbonylferrate 四羧基铁酸二钠
16. Cyclopentadienyliron dicarbonyl dimer 双(二羰基环戊二烯铁)
17. catalyst 催化剂
18. transfer hydrogenation 转移氢化反应
19. ketone 酮
20. Mohr's salt 莫尔盐

Exercises

1. Describe iron's origin and occurrence in nature.
2. What are organometallic compounds?
3. What are iron's industrial uses?

Passage E Etymology

As iron has been in use for such a long time, it has many names. The source of its chemical symbol Fe is the Latin word *ferrum*, and its descendants are the names of the element in the Romance languages (for example, French *fer*, Spanish *hierro*, and Italian and Portuguese *ferro*). The word "ferrum" itself possibly comes from the Semitic languages, via Etruscan, from a root that also gave rise to Old English *bræs* "brass". The English word iron derives ultimately from Proto-Germanic *isarnan*, which is also the source of the German name *Eisen*. It was most likely borrowed from Celtic *isarnon*, which ultimately comes from Proto-Indo-European *is-(e)ro-* "powerful, holy" and finally *eis* "strong", referencing iron's strength as a metal. Kluge relates *isarnon* to Illyric and Latin *ira* "wrath"). The Balto-Slavic names for iron (e. g. Russian *железо*, Polish *żelazo*, Lithuanian *geležis*) are the only ones to come directly from the Proto-Indo-European *ghelgh-* "iron". In many of these languages, the word for iron may also be used to denote other objects made of iron or steel, or figuratively because of the hardness and strength of the metal. The Chinese *tiě* (traditional 鐵; simplified 铁) derives from Proto-Sino-Tibetan *hliek*, and was borrowed into Japanese as 鉄 (*tetsu*), which also has the native reading *kurogane* "lack metal" (similar to how iron is referenced in the English word blacksmith).

Terms

blacksmith 铁匠

Exercises

1. What is the source of iron's chemical symbol Fe?
2. Where does the word *ferrum* possibly come from?
3. Where does the Chinese *tiě* (traditional 鐵; simplified 铁) derive from?

Passage F History

Development of iron metallurgy

Iron is one of the elements undoubtedly known to the ancient world. It has been worked, or wrought, for millennia. However, iron objects of great age are much rarer than objects made of gold or silver due to the ease with which iron corrodes. The technology developed slowly, and even after the discovery of smelting, it took many centuries for iron to replace bronze as the metal of choice for tools and weapons.

Meteoritic iron

Beads made from meteoric iron in 3,500 BC or earlier were found in Gerzah, Egypt by G. A. Wainwright. The beads contain 7.5% nickel, which is a signature of meteoric origin since iron found in the Earth's crust generally has only minuscule nickel impurities.

Meteoric iron was highly regarded due to its origin in the heavens and was often used to forge weapons and tools. For example, a dagger made of meteoric iron was found in the tomb of Tutankhamun, containing similar proportions of iron, cobalt, and nickel to a meteorite discovered in the area, deposited by an ancient meteor shower. Items that were likely made of iron by Egyptians date from 3,000 to 2,500 BC.

Meteoritic iron is comparably soft and ductile and easily cold-forged but may get brittle when heated because of the nickel content.

Wrought iron

The first iron production started in the Middle Bronze Age, but it took several centuries before iron displaced bronze. Samples of smelted iron from Asmar, Mesopotamia, and Tall Chagar Bazaar in northern Syria were made sometime between 3,000 and 2,700 BC. The Hittites established an empire in northcentral Anatolia around 1,600 BC. They appear to be the first to understand the production of iron from its ores and regard it highly in their society. The Hittites began to smelt iron between 1,500 and 1,200 BC and the practice spread to the rest of the Near East after their empire fell in 1,180 BC. The subsequent period is called the Iron Age.

Artifacts of smelted iron are found in India dating from 1,800 to 1,200 BC, and in the Levant from about 1,500 BC (suggesting smelting in Anatolia or the Caucasus). Alleged references (compare the history of metallurgy in South Asia) to iron in the Indian Vedas have been used for claims of a very early usage of iron in India respectively to date the texts as such. The rigveda term ayas (metal) probably refers to copper and bronze, while iron or sydmaayas, literally "black metal", first is mentioned in the post-rigvedic Atharvaveda.

Some archaeological evidence suggests iron was smelted in Zimbabwe and southeast Africa as early as the 8th century BC. Ironworking was introduced to Greece in the late 11th century BC, from which it spread quickly throughout Europe.

The spread of ironworking in Central and Western Europe is associated with Celtic expansion. According to Pliny the Elder, iron use was common in the Roman era. The annual iron output of the Roman Empire is estimated 84,750 t. The use of a blast furnace in China dates to the 1st century AD, and cupola furnaces were used as early as the Warring States period (403 - 221 BC). Usage of the blast and cupola furnace remained widespread during the Song and Tang Dynasties.

During the Industrial Revolution in Britain, Henry Cort began refining iron from pig iron to wrought iron (or bar iron) using innovative production systems. In 1783 he patented the puddling process for refining iron ore. It was later improved by others, including Joseph Hall.

Cast iron

Cast iron was first produced in China during the 5th century BC, but was hardly in Europe until the medieval period. The earliest cast iron artifacts were discovered by archaeologists in what is now modern Luhe County, Jiangsu in China. Cast iron was used in ancient China for warfare, agriculture and architecture. During the medieval period, means were found in Europe of producing wrought iron from cast iron (in this context known as pig iron) using finery forges. For all these processes, charcoal was required as fuel.

Medieval blast furnaces were about 10 feet tall and made of fireproof brick; forced air was usually provided by hand-operated bellows. Modern blast furnaces have grown much bigger, with hearths fourteen meters in diameter that allow them to produce thousands of tons of iron each day, but essentially operate in much the same way as they did during medieval times.

In 1709, Abraham Darby I established a coke-fired blast furnace to produce cast iron, replacing charcoal, although continuing to use blast furnaces. The ensuing availability of inexpensive iron was one of the factors leading to the Industrial Revolution. Toward the end of the 18th century, cast iron began to replace wrought iron for certain purposes, because it was cheaper. Carbon content in iron was not implicated as the reason for the differences in properties of wrought iron, cast iron, and steel until the 18th century.

Since iron was becoming cheaper and more plentiful, it also became a major structural material following the building of the innovative first iron bridge in 1778. This bridge still stands today as a monument to the role iron played in the Industrial Revolution. Following this, iron was used in rails, boats, ships, aqueducts, and buildings, as well as in iron cylinders in steam engines. Railways have been central to the formation of modernity and ideas of progress, and various languages (e. g. French, Spanish, Italian, and German) refer to railways as iron "roads".

Steel

Steel (with smaller carbon content than pig iron but more than wrought iron) was first produced in antiquity by using bloomery. Blacksmiths in Luristan in western Persia were making good steel by 1,000 BC. Then improved versions, Wootz steel by India and Damascus steel were developed around 300 BC and 500 AD, respectively. These methods were specialized, and steel did not become a major commodity until the 1850s.

New methods of producing it by carburizing bars of iron in the cementation process were devised in the 17th century. In the Industrial Revolution, new methods of producing bar iron without charcoal were devised and these were later applied to produce steel. In the late 1850s, Henry Bessemer invented a new steelmaking process, involving blowing air through molten pig iron, to produce mild steel. This made steel much more economical, thereby leading to wrought iron no longer being produced in large quantities.

Foundations of modern chemistry

In 1774, Antoine Lavoisier used the reaction of water steam with metallic iron inside an incandescent iron tube to produce hydrogen in his experiments leading to the demonstration of the conservation of mass, which was instrumental in changing chemistry from a qualitative science to a quantitative one.

Terms

1. wrought 锻造的;制作的
2. dagger 匕首;短剑
3. meteorite 陨石
4. cold forged 冷锻的
5. brittle 易碎的
6. Middle Bronze Age 中青铜时代
7. blast furnace 鼓风炉
8. pig iron 生铁
9. cast iron 铸铁,生铁
10. medieval 中古的,中世纪的
11. charcoal 木炭
12. the Industrial Revolution 工业革命
13. steam engine 蒸汽机;蒸馏机
14. Wootz steel 乌兹钢
15. carburize 使渗碳;使与碳结合
16. cementation process 渗碳法
17. hydrogen 氢
18. conservation of mass 质量守恒定律
19. puddling process 普德林法(搅炼法)
20. Damascus steel 大马士革钢

Exercises

1. Briefly explain the development of iron and steel metallurgy.
2. What did Abraham Darby I establish?
3. How was steel first produced?

Passage G Symbolic Role

Iron plays a certain role in mythology and has found various usage as a metaphor and in folklore. The Greek poet Hesiod's *Works and Days* (Lines 109 - 201) lists different ages of man named after metals like gold, silver, bronze and iron to account for successive ages of humanity. The Iron Age was closely related with Rome, and in Ovid's Metamorphoses:

> The Virtues, in despair, quit the earth; and the depravity of man becomes universal and complete. Hard steel succeeded then.
>
> —Ovid, Metamorphoses, *Book* I, Iron Age, line 160 ff

An example of the importance of iron's symbolic role may be found in the German Campaign of 1813. Frederick William III commissioned then the first Iron Cross as military decoration. Berlin iron jewellery reached its peak production between 1813 and 1815 when the Prussian royal family urged citizens to donate gold and silver jewellery for military funding. The inscription Gold gab ich fur Eisen (I gave gold for iron) was used as well in later war efforts.

Exercises

1. What do you think of metals like gold, silver, bronze and iron in the Greek poet Hesiod's *Works and Days*?
2. Which city's iron jewellery reached its peak production between 1813 and 1815?
3. Whose work is the quoted prose?

Passage H Production of Metallic Iron

Laboratory routes

For a few limited purposes when it is needed, pure iron is produced in the laboratory in small quantities by reducing the pure oxide or hydroxide with hydrogen, or forming iron pentacarbonyl and heating it to 250 ℃ so that it decomposes to form pure iron powder. Another method is the electrolysis of ferrous chloride onto an iron cathode.

Main industrial route

Nowadays, the industrial production of iron or steel consists of two main stages. In the first stage, iron ore is reduced with coke in a blast furnace, and the molten metal is separated from gross impurities such as silicate minerals. This stage yields an alloy—pig iron—that contains relatively large amounts of carbon. In the second stage, the amount of carbon in the pig iron is lowered by oxidation to yield wrought iron, steel, or cast iron. Other metals can be added at this stage to form alloy steels.

Blast furnace processing

The blast furnace is loaded with iron ores, usually hematite Fe_2O_3 or magnetite Fe_3O_4, together with coke (coal that has been separately baked to remove volatile components). Air pre-heated to 900 ℃ is blown through the mixture, in sufficient amount to turn the carbon into carbon monoxide:

$$2C + O_2 \rightarrow 2CO$$

This reaction raises the temperature to about 2,000 ℃. The carbon monoxide reduces the iron ore to metallic iron:

$$Fe_2O_3 + 3CO \rightarrow 2Fe + 3CO_2$$

Some iron in the high-temperature lower region of the furnace reacts directly with the coke:

$$2Fe_2O_3 + 3C \rightarrow 4Fe + 3CO_2$$

A flux such as limestone (calcium carbonate) or dolomite (calcium-magnesium carbonate) is also added to the furnace's load. Its purpose is to remove silicaceous minerals in the ore, which would otherwise clog the furnace. The heat of the furnace decomposes the carbonates to calcium oxide, which reacts with any excess silica to form a slag composed of calcium silicate $CaSiO_3$ or other products. At the furnace's temperature, the metal and the slag are both molten. They collect at the bottom as two immiscible liquid layers (with the slag on top), which are then easily separated. The slag can be used as a material in road construction or to improve mineral-poor soils for agriculture.

Steelmaking

In general, the pig iron produced by the blast furnace process contains up to 4%-5% carbon, with small amounts of other impurities like sulfur, magnesium, phosphorus, and manganese. The high level of carbon makes it relatively weak and brittle. Reducing the amount of carbon to 0.002% – 2.1% by mass produces steel, which may be up to 1,000 times harder than pure iron. A great variety of steel articles can then be made by cold working, hot rolling, forging, machining, etc. Removing the other impurities, instead, results in cast iron, which is used to cast articles in foundries, for example, stoves, pipes, radiators, lamp-posts, and rails.

Steel products often undergo various heat treatments after they are forged to shape. Annealing consists of heating them to 700 ℃ – 800 ℃ for several hours and then gradual cooling. It makes the steel softer and more workable.

Direct iron reduction

Owing to environmental concerns, alternative methods of making steel processing iron have been developed. "Direct iron reduction" reduces iron ore to a ferrous lump called "sponge" iron or "direct" iron that is suitable for steelmaking. Two main reactions comprise the direct reduction process:

Natural gas is partially oxidised (with heat and a catalyst):

$$2CH_4 + O_2 \rightarrow 2CO + 4H_2$$

Iron ore is then treated with these gases in a furnace, producing solid sponge iron:

$$Fe_2O_3 + CO + 2H_2 \rightarrow 2Fe + CO_2 + 2H_2O$$

Silica is removed by adding a limestone flux as described above.

Thermite process

Ignition of a mixture of aluminium powder and iron oxide yields metallic iron via the thermite reaction:

$$Fe_2O_3 + 2Al \rightarrow 2Fe + Al_2O_3$$

Alternatively, pig iron may be made into steel (with up to about 2% carbon) or wrought iron (commercially pure iron). Various processes have been used for this, including finery forges, puddling furnaces, Bessemer converters, open hearth furnaces, basic oxygen furnaces, and electric arc furnaces. In all cases, the objective is to oxidise some or all of the carbon, together with other impurities. On the other hand, other metals may be added to make alloy steels.

Terms

1. silicate mineral 硅酸盐矿物	2. limestone 石灰岩
3. calcium carbonate 碳酸钙	4. dolomite 白云石
5. calcium oxide 氧化钙	6. slag 矿渣;熔渣
7. calcium silicate 硅酸钙	8. hot rolling 热轧
9. machine 用机器切割,制作或磨光(某物)	10. foundry 铸造厂
11. heat treatment 热处理	12. sponge iron 海绵铁
13. direct iron 直接还原铁	14. thermite reaction 热石反应
15. finery forge 精炼	16. puddling furnace 搅炼炉
17. Bessemer converter 酸性转炉	18. open hearth furnace 平炉
19. electric arc furnace 电弧炉	20. anneal 退火

Exercises

1. What consists of two main stages of the industrial production of iron or steel nowadays?

2. What is the blast furnace's purpose?

3. In general, how much carbon is contained in the pig iron produced by the blast furnace process?

Passage I Applications

As structural material

Iron is the most widely used of all metals, accounting for over 90% of worldwide metal production. Its low cost and high strength often make it the material of choice to withstand stress or transmit forces, such as the construction of machinery and machine tools, rails, automobiles, ship hulls, concrete reinforcing bars, and the load-carrying framework of buildings. Since pure iron is quite soft, it is most commonly combined with alloying elements to make steel.

Mechanical properties

The mechanical properties of iron and its alloys are extremely relevant to their structural applications. Those properties can be evaluated in various ways, including the Brinell test, the Rockwell test and the Vickers hardness test. The properties of pure iron are often used to calibrate measurements or to compare tests. However, the mechanical properties of iron are significantly affected by the sample's purity: pure, single crystals of iron are actually softer than, and the purest industrially produced iron (99.99%) has a hardness of 20 – 30 Brinell. An increase in the carbon content will cause a significant increase in the hardness and tensile strength of iron. Maximum hardness of 65 Rc is achieved with a 0.6% carbon content, although the alloy has low tensile strength. Because of the softness of iron, it is much easier to work with than its heavier congeners ruthenium and osmium.

Types of steels and alloys

α-iron is a fairly soft metal that can dissolve only a small concentration of carbon (no more than 0.021% by mass at 910 ℃). Austenite (γ-iron) is similarly soft and metallic but can dissolve considerably more carbon (as much as 2.04% by mass at 1,146 ℃). This form of iron is used in the type of stainless steel for making cutlery, and hospital and food-service equipment.

Commercially available iron is classified based on purity and abundance of additives. Pig iron has 3.5% – 4.5% carbon and contains varying amounts of contaminants such as sulfur, silicon and phosphorus. Pig iron is not a saleable product, but rather an intermediate step in the production of cast iron and steel. The reduction of contaminants in pig iron that negatively affect material properties, such as sulfur and phosphorus, yields cast iron containing 2% – 4% carbon, 1% – 6% silicon, and small amounts of manganese. Pig iron has an iron-carbon phase diagram melting point in the range of 1,420 – 1,470 K, which is lower than either of its two main components and makes it the first product to be melted when carbon and iron are heated together. Its mechanical properties vary greatly and depend on the form the carbon takes in the alloy.

"White" cast irons contain their carbon in the form of cementite, or iron carbide (Fe_3C). This hard, brittle compound dominates the mechanical properties of white cast irons, rendering them hard, but irresistant unresistant to shock. The broken surface of white cast iron is full of fine facets of the broken iron carbide, a very pale, silvery, shiny material, hence the appellation. Cooling a mixture of iron with 0.8% carbon slowly below 723 ℃ to room temperature results in separate, alternating layers of cementite and α-iron, which is soft and malleable and is called pearlite for its appearance. Rapid cooling, on the other hand, does not allow time for this separation and creates hard and brittle martensite. The steel can then be tempered by reheating to a temperature in between, changing the proportions of pearlite and martensite. The end product below 0.8% carbon content is a pearlite-αFe mixture, and that above 0.8% carbon content is a pearlite-cementite mixture.

In gray iron the carbon exists as separate, fine flakes of graphite, and also renders the material brittle due to the sharp-edged flakes of graphite that produce stress concentration sites within the material. A newer variant of gray iron, referred to as ductile iron, is specially treated with trace amounts of magnesium to alter the shape of graphite to spheroids, or nodules, reducing the stress concentrations and vastly increasing the toughness and strength of the material.

Wrought iron contains less than 0.25% carbon but large amounts of slag that give it a fibrous characteristic. It is a tough, malleable product, but not as fusible as pig iron. Wrought iron is characterized by the presence of fine fibers of slag entrapped within the metal. Wrought iron is more corrosion-resistant than steel. It has been almost completely replaced by mild steel for traditional "wrought iron" products and blacksmithing.

Mild steel corrodes more readily than wrought iron but is cheaper and more widely available. Carbon steel contains 2.0% carbon or less, with small amounts of manganese, sulfur, phosphorus, and silicon. Alloy steels contain varying amounts of carbon as well as other metals, such as chromium, vanadium, molybdenum, nickel, tungsten. Their alloy content raises their cost, and so they are usually only employed for specialist uses. One common alloy steel, though, is stainless steel. Recent developments in ferrous metallurgy have produced a growing range of microalloyed steels, also termed "HSLA" or high-strength, low-alloy steels, containing tiny additions to produce high strengths and often spectacular toughness at a minimal cost.

Apart from traditional applications, iron is also used for protection from ionizing radiation. Although it is lighter than another traditional protection material, lead, it is much stronger mechanically.

The main disadvantage of iron and steel is that pure iron, and most of its alloys, suffer badly from rust if not protected in some way, a cost amounting to over 1% of the world's economy. Painting, galvanization, passivation, plastic coating and bluing are all used to protect iron from rust by excluding water and oxygen or by cathodic protection. The mechanism of the rusting of iron is as follows:

Cathode: $3O_2 + 6H_2O + 12e^- \rightarrow 12OH^-$

Anode: $4Fe \rightarrow 4Fe^{2+} + 8e^-$; $4Fe^{2+} \rightarrow 4Fe^{3+} + 4e^-$

Overall: $4Fe + 3O_2 + 6H_2O \rightarrow 4Fe^{3+} + 12OH^- \rightarrow 4Fe(OH)_3/4FeO(OH) + 4H_2O$

The electrolyte is usually iron (II) sulfate in urban areas (formed when atmospheric sulfur dioxide attacks iron), and salt particles in the atmosphere in seaside areas.

Iron compounds

Although the dominant use of iron is in metallurgy, iron compounds are also pervasive in the industry. Iron catalysts are traditionally used in the Haber-Bosch process for the production of ammonia and the Fischer-Tropsch process for the conversion of carbon monoxide to hydrocarbons for fuels and lubricants. Powdered iron in an acidic solvent was used in the Bechamp reduction of nitrobenzene to aniline.

Iron (III) oxide mixed with aluminium powder can be ignited to create a thermite reaction, used in welding large iron parts (like rails) and purifying ores. Iron (III) oxide and oxyhidroxide are used as reddish and ocher pigments.

Iron (III) chloride finds use in water purification and sewage treatment, in the dyeing of cloth, as a coloring agent in paints, as an additive in animal feed, and as an etchant for copper in the manufacture of printed circuit boards. It can also be dissolved in alcohol to form tincture of iron, which is used as a medicine to stop bleeding in canaries.

Iron (II) sulfate is used as a precursor to other iron compounds. It is also used to reduce chromate in cement. It is used to fortify foods and treat iron deficiency anemia. Iron (III) sulfate is used in settling minute sewage particles in tank water. Iron (II) chloride is used as a reducing flocculating agent, in the formation of iron complexes and magnetic iron oxides, and as a reducing agent in organic synthesis.

Terms

1. ship hull 船体
2. concrete reinforcing bar 混凝土钢筋
3. Brinell test 布氏试验,布氏硬度试验
4. Rockwell test 洛氏硬度试验
5. Vickers hardness test 维氏硬度试验
6. sulfur 硫黄
7. phosphorus 磷;红磷
8. cementite 渗碳体
9. pearlite 珠光体
10. martensite 马氏体
11. gray iron 灰口铁,灰口铸铁,灰铸铁
12. graphite 石墨
13. stress concentration 应力集中
14. wrought iron 熟铁,锻铁
15. mild steel 低碳钢
16. carbon steel 碳钢
17. vanadium 钒
18. molybdenum 钼

19. tungsten 钨
20. passivation 钝化
21. hydrocarbon 碳氢化合物
22. nitrobenzene 硝基苯
23. aniline 苯胺
24. printed circuit board 印制电路板
25. iron deficiency anemia 缺铁性贫血
26. cathodic protection 阴极保护
27. etchant (用于金属、玻璃等的)蚀刻剂
28. Fischer-Tropsch process 费托法

Exercises

1. What types of steels and alloys are there?
2. How many iron compounds are there in nature?
3. What is the iron sulfate used?

Passage J Biological and Pathological Role

Iron is required for life. The iron-sulfur clusters are pervasive and include nitrogenase, the enzymes responsible for biological nitrogen fixation. Iron-containing proteins participate in the transport, storage and use of oxygen. Iron proteins are involved in electron transfer.

Examples of iron-containing proteins in higher organisms include hemoglobin, cytochrome (see high-valent iron), and catalase. The average adult human contains about 0.005% body weight of iron, or about four grams, of which three quarters is in hemoglobin—a level that remains constant despite only about one milligram of iron being absorbed each day because the human body recycles its hemoglobin for the iron content.

Biochemistry

Iron acquisition poses a problem for aerobic organisms because ferric iron is poorly soluble near neutral pH. Thus, these organisms have developed means to absorb iron as complexes, sometimes taking up ferrous iron before oxidizing it back to ferric iron. In particular, bacteria have evolved very high-affinity sequestering agents called siderophores.

After uptake in human cells, iron storage is precisely regulated. A major component of this regulation is the protein transferrin, which binds iron ions absorbed from the duodenum and carries it in the blood to cells. Transferrin contains Fe^{3+} in the middle of a distorted octahedron, bonded to one nitrogen, three oxygens and a chelating carbonate anion that traps the Fe^{3+} ion: it has such a high stability constant that it is very effective at taking up Fe^{3+} ions even from the most stable complexes. At the bone marrow, transferrin is reduced from Fe^{3+} and Fe^{2+} and stored as ferritin to be incorporated into hemoglobin.

The most commonly known and studied bioinorganic iron compounds (biological iron molecules) are heme proteins: examples are hemoglobin, myoglobin, and cytochrome P_{450}. These compounds participate in transporting gases, building enzymes, and transferring electrons. Metalloproteins are a group of proteins with metal ion cofactors. Some examples of iron metalloproteins are ferritin and rubredoxin. Many enzymes vital to life contain iron, such as catalase, lipoxygenases, and IRE-BP.

Hemoglobin is an oxygen carrier that occurs in red blood cells and contributes their color, transporting oxygen in the arteries from the lungs to the muscles where it is transferred to myoglobin, which stores it until it is needed for the metabolic oxidation of glucose, generating energy. Here the hemoglobin binds to carbon dioxide, produced when glucose is oxidised, which is transported through the veins by hemoglobin (predominantly as bicarbonate anions) back to the lungs where it is exhaled. In hemoglobin, the iron is in one of four heme groups and has six possible coordination sites; four are occupied by nitrogen atoms in a porphyrin ring, the fifth by imidazole nitrogen in a histidine residue of one of the protein chains attached to the heme group, and the sixth is reserved for the oxygen molecule it can reversibly bind to. When hemoglobin is not attached to oxygen (and is then called deoxyhemoglobin), the Fe^{2+} ion at the center of the heme group (in the hydrophobic protein interior) is in a high-spin configuration. It is thus too large to fit inside the porphyrin ring, which bends instead into a dome with the Fe^{2+} ion about 55 picometers above it. In this configuration, the sixth coordination site reserved for the oxygen is blocked by another histidine residue.

When deoxyhemoglobin picks up an oxygen molecule, this histidine residue moves away and returns once the oxygen is securely attached to form a hydrogen bond with it. This results in the Fe^{2+} ion switching to a low-spin configuration, resulting in a 20% decrease in ionic radius so that now it can fit into the porphyrin ring, which becomes planar. (Additionally, this hydrogen bonding results in the tilting of the oxygen molecule, resulting in a Fe-O-O bond angle of around 120o that avoids the formation of Fe-O-Fe or Fe-O_2-Fe bridges that would lead to electron transfer, the oxidation of Fe^{2+} to Fe^{3+}, and the destruction of hemoglobin.) This results in a movement of all the protein chains that leads to the other subunits of hemoglobin changing shape to a form with larger oxygen affinity. Thus, when deoxyhemoglobin takes up oxygen, its affinity for more oxygen increases, and vice versa. Myoglobin, on the other hand, contains only one heme group and hence this cooperative effect cannot occur. Thus, while hemoglobin is almost saturated with oxygen in the high partial pressures of oxygen found in the lungs, its affinity for oxygen is much lower than that of myoglobin, which oxygenates even at low partial pressures of oxygen found in muscle tissue. As described by the Bohr effect (named after Christian Bohr, the father of Niels Bohr), the oxygen affinity of hemoglobin diminishes in the presence of carbon dioxide.

Carbon monoxide and phosphorus trifluoride are poisonous to humans because they bind to hemoglobin similarly to oxygen, but with much more strength, so that oxygen can no longer be transported throughout the body. Hemoglobin bound to carbon monoxide is known as carboxyhemoglobin. This effect also plays a minor role in the toxicity of cyanide, but there the major effect is by far its interference with the proper functioning of the electron transport protein cytochrome. The cytochrome proteins also involve heme groups and are involved in the metabolic oxidation of glucose by oxygen. The sixth coordination site is then occupied by either another imidazole nitrogen or methionine sulfur so that these proteins are largely inert to oxygen—except for cytochrome, which bonds directly to oxygen and thus is very easily poisoned by cyanide. Here, the electron transfer takes place as the iron remains in a low spin but changes between the +2 and +3 oxidation states. Since the reduction potential of each step is slightly greater than the previous one, the energy is released step-by-step and can thus be stored in adenosine triphosphate. Cytochrome a is slightly distinct, as follows:

$$4Cytc^{2+} + O_2 + 8H^+_{inside} \rightarrow 4Cytc^{3+} + H_2O + 4H^+_{outside}$$

Although the heme proteins are the most important class of iron-containing proteins, the iron sulfur proteins are also very important, being involved in electron transfer, which is possible since iron can exist stably in either the +2 or +3 oxidation states. These have one, two, four, or eight iron atoms that are each approximately tetrahedrally coordinated to four sulfur atoms; because of this tetrahedral coordination, they always have high-spin iron. The simplest of such compounds is rubredoxin, which has only one iron atom coordinated to four sulfur atoms from cysteine residues in the surrounding peptide chains. Another important class of iron-sulfur proteins is the ferredoxins, which have multiple iron atoms. Transferrin does not belong to either of these classes.

The ability of sea mussels to maintain their grip on rocks in the ocean is facilitated by their use of organometallic iron-based bonds in their protein-rich cuticles. Based on synthetic replicas, the presence of iron in these structures increased elastic modulus 770 times, tensile strength 58 times, and toughness 92 times. The amount of stress required to permanently damage them increased 76 times.

Nutrition diet

Iron is pervasive, and particularly rich sources of dietary iron include red meat, oysters, lentils, beans, poultry, fish, leaf vegetables, watercress, tofu, chickpeas, black-eyed peas, and blackstrap molasses. Bread and breakfast cereals are sometimes specifically fortified with iron.

Iron provided by dietary supplements is often found as iron (II) fumarate, although iron (II) sulfate is cheaper and is absorbed equally well. Elemental iron, or reduced iron, despite being absorbed in only one-third to two-thirds the efficiency (relative to iron sulfate), is often added to foods such as breakfast cereals or enriched wheat flour. Iron is most available to the body when chelated to amino acids and is also available for use as a common iron supplement. Glycine, the least expensive amino acid, is most often used to produce iron glycinate supplements.

Dietary recommendations

The US Institute of Medicine (IOM) updated Estimated Average Requirements (EARs) and Recommended Dietary Allowances (RDAs) for iron in 2001. The current EAR for iron for women ages 14 - 18 is 7.9 mg/day, 8.1 for ages 19 - 50 and 5.0 thereafter (post menopause). For men the EAR is 6.0 mg/day for ages 19 and up. The RDA is 15.0 mg/day for women ages 15 - 18, 18.0 for 19 - 50 and 8.0 thereafter. For men, 8.0 mg/day for ages 19 and up. RDAs are higher than EARs to identify amounts that will cover people with higher than average requirements. RDA for pregnancy is 27 mg/day and, for lactation, 9 mg/day. For children ages 1 - 3 years 7 mg/day, 10 for ages 4 - 8 and 8 for ages 9 - 13. As for safety, the IOM also sets tolerable upper intake levels (UL) for vitamins and minerals when evidence is sufficient. In the case of iron, the UL is set at 45 mg/day. Collectively the EARs, RDAs and ULs are referred to as Dietary Reference Intake (DRI).

The European Food Safety Authority (EFSA) refers to the collective set of information as Dietary Reference Values, with Population Reference Intake (PRI) instead of RDA, and Average Requirement instead of EAR. AI and UL are defined the same as in the United States. For women, the PRI is 13 mg/day ages 15 - 17 years, 16 mg/day for women ages 18 and up who are premenopausal and 11 mg/day postmenopausal. For pregnancy and lactation, 16 mg/day. For men, the PRI is 11 mg/day ages 15 and older. For children ages 1 to 14, the PRI increases from 7 to 11 mg/day. The PRIs are higher than the US RDAs, except for pregnancy. The EFSA reviewed the same safety question did not establish a UL.

Infants may require iron supplements if they are bottle-fed cow's milk. Frequent blood donors are at risk of low iron levels and are often advised to supplement their iron intake.

For US food and dietary supplement labeling purposes the amount in a serving is expressed as a percent of Daily Value (% DV). For iron labeling purposes 100% of the Daily Value was 18 mg, and as of May 27, 2016 remained unchanged at 18 mg. A table of all of the old and new adult Daily Values is provided at Reference Daily Intake. The original deadline to comply was July 28, 2018, but on September 29, 2017, the US Food and Drug Administration released a proposed rule that extended the deadline to January 1, 2020 for large companies and January 1, 2021 for small companies.

Deficiency

Iron deficiency is the most common nutritional deficiency in the world. When loss of iron is not adequately compensated by adequate dietary iron intake, a state of latent iron deficiency occurs, which over time leads to iron-deficiency anemia if left untreated, which is characterized by an insufficient number of red blood cells and an insufficient amount of hemoglobin. Children, pre-menopausal women (women of child-bearing age), and people with poor diets are most susceptible to the disease. Most cases of iron-deficiency anemia are mild, but if not treated can cause problems like fast or irregular heartbeat, complications during pregnancy, and delayed growth in infants and children.

Excess

Iron uptake is tightly regulated by the human body, which has no regulated physiological means of excreting iron. Only small amounts of iron are lost daily due to mucosal and skin epithelial cell sloughing, so control of iron levels is primarily accomplished by regulating uptake. Regulation of iron uptake is impaired in some people as a result of a genetic defect that maps to the HLA-H gene region on chromosome 6 and leads to abnormally low levels of hepcidin, a key regulator of the entry of iron into the circulatory system in mammals. In these people, excessive iron intake can result in iron overload disorders, known medically as hemochromatosis. Many people have an undiagnosed genetic susceptibility to iron overload, and are not aware of a family history of the problem. For this reason, people should not take iron supplements unless they suffer from iron deficiency and have consulted a doctor. Hemochromatosis is estimated to be the cause of 0.3% to 0.8% of all metabolic diseases of Caucasians.

Overdoses of ingested iron can cause excessive levels of free iron in the blood. High blood levels of free ferrous iron react with peroxides to produce highly reactive free radicals that can damage DNA, proteins, lipids, and other cellular components. Iron toxicity occurs when the cell contains free iron, which generally occurs when iron levels exceed the availability of transferrin to bind the iron. Damage to the cells of the gastrointestinal tract can also prevent them from regulating iron absorption, leading to further increases in blood levels. Iron typically damages cells in the heart, liver and elsewhere, causing adverse effects that include coma, metabolic acidosis, shock, liver failure, coagulopathy, adult respiratory distress syndrome, long-term organ damage, and even death. Humans experience iron toxicity when the iron exceeds 20 milligrams for every kilogram of body mass; 60 milligrams per kilogram is considered a lethal dose. Over-consumption of iron, often the result of children eating large quantities of ferrous sulfate tablets intended for adult consumption, is one of the most common toxicological causes of death in children under six. The DRI sets the UL for adults at 45 mg/day. For children under fourteen years old, the UL is 40 mg/day.

The medical management of iron toxicity is complicated, and can include the use of a specific chelating agent called deferoxamine to bind and expel excess iron from the body.

Cancer

The role of iron in cancer defense can be described as a "double-edged sword" because of its pervasive presence in non-pathological processes. People having chemotherapy may develop iron deficiency and anemia, for which intravenous iron therapy is used to restore iron levels. Iron overload, which may occur from high consumption of red meat, may initiate tumor growth and increase susceptibility to cancer onset, particularly for colorectal cancer.

Terms

1. iron-sulfur cluster 铁硫簇
2. nitrogen fixation 固氮(作用)
3. electron transfer 电子传递,电子转移
4. cytochrome 细胞色素
5. high-valent iron 高价铁
6. catalase 过氧化氢酶
7. sequester 使隔绝;使隔离
8. siderophore 铁载体
9. transferrin 铁传递蛋白
10. duodenum 十二指肠
11. ferritin 铁蛋白
12. bioinorganic 生物无机(化学)的
13. heme protein 血红素蛋白
14. metalloprotein 金属蛋白
15. cofactor 辅酶因子
16. lipoxygenases 脂氧合酶
17. glucose 葡萄糖;右旋糖
18. bicarbonate 碳酸氢盐,重碳酸盐
19. porphyrin 卟啉
20. imidazole 咪唑
21. histidine 组氨酸
22. heme 血红素
23. hydrogen bond 氢键
24. Bohr effect 玻尔效应
25. Christian Bohr 克里斯蒂安·波尔
26. Niels Bohr 尼尔斯·波尔
27. phosphorus trifluoride 三氟化磷
28. carboxyhemoglobin 碳氧血红蛋白
29. methionine 蛋氨酸,甲硫氨酸
30. adenosine triphosphate 三磷酸腺苷
31. sulfur protein 硫蛋白
32. tetrahedrally 四面体地
33. glycine 甘氨酸
34. the US Institute of Medicine (IOM) 美国医学院
35. Estimated Average Requirements (EARs) 平均需要量
36. Recommended Dietary Allowances (RDAs) 推荐的膳食中营养素供给量
37. pathological 病理学的;由疾病引起的;病态的,疾病的
38. rubredoxin 红素氧还蛋白,红氧(化)还(原)蛋白
39. IRE-BP (iron responsive element-binding protein)铁反应子结合蛋白
40. tolerable upper intake level 可耐受的上摄入量水平
41. Dietary Reference Intake 膳食参考摄入量
42. European Food Safety Authority 欧洲食品安全管理局
43. Population Reference Intake 人群参考摄入量
44. the US Food and Drug Administration 美国食品药品监督管理局
45. adult respiratory distress syndrome 成人型呼吸窘迫综合征

Exercises

1. Why is iron required for life?

2. Why can the role of iron in cancer defense be described as a "double-edged sword"?

2. Steel

Passage A Introduction

It is well-known that iron and steel are the most useful materials in the world. Compared with other materials, iron and steel are more popular. They rank second in the Earth's crust, only next to that of aluminum. The smelting of iron ore is easier than other elements. It is featured by manufacturing, high efficiency, and low cost. Iron and steel also can replace other elements in certain applications, due to its metallic features of hardness and strength.

Steel is an alloy of iron and carbon, and sometimes other elements like chromium. Because of its high tensile strength and low cost, it is a major component used in buildings, infrastructure, tools, ships, trains, automobiles, machines, appliances, and weapons.

Iron is the base metal of steel. Iron can take on two crystalline forms (allotropic forms), body-centered cubic and face-centered cubic, depending on its temperature. In the body-centered cubic arrangement, there is an iron atom in the center and eight atoms at the vertices of each cubic unit cell; in the face-centered cubic, there is one atom at the center of each of the six faces of the cubic unit cell and eight atoms at its vertices. It is the interaction of the allotropes of iron with the alloying elements, primarily carbon, which gives steel and cast iron their range of unique properties.

In pure iron, the crystal structure has relatively little resistance to the iron atoms slipping past one another, and so pure iron is quite ductile, or soft and easily formed. In steel, small amounts of carbon, other elements, and inclusions within the iron act as hardening agents that prevent the movement of dislocations.

The carbon in typical steel alloys may contribute up to 2.14% of its weight. Varying the amount of carbon and many other alloying elements, as well as controlling their chemical and physical makeup in the final steel (either as solute elements, or as precipitated phases), slows the movement of those dislocations that make pure iron ductile, and thus controls and enhances its qualities. These qualities include the hardness, quenching behavior, need for annealing, tempering behavior, yield strength, and tensile strength of the resulting steel. The increase in steel's strength compared to pure iron is possible only by reducing iron's ductility.

Although steel and iron are both iron-carbon alloys, the difference of carbon and other elements, especially the carbon content, will lead to the changes of properties and structures at different temperatures. As a result, iron and steel are totally different. The amount of carbon content is the main standard to distinguish steel and iron. In general, iron with a carbon content of more than 2% is called pig iron, and steel is with a carbon content of less than 2% (Table 2-1).

Table 2-1 Elements of Pig Iron and Steel

Elements	C/%	Si/%	Mn/%	P/%	S/%
Pig iron	>2	0.2 - 2.0	0.2 - 2.5	≤0.5	≤7
Steel	≤2	0.01 - 0.3	0.3 - 0.8	<0.05	<0.05

As time goes by, steel is becoming a more widely used material in our daily life. In terms of its production, there are three major advantages. The supply of its raw material is rich (included iron-bearing material, flux, and flue). And the production of single equipment is developing fast. It requires less and less material as the high technology is developing.

Steel was produced in bloomery furnaces for thousands of years, but its large-scale, industrial use began only after more efficient production methods were devised in the 17th century, with the introduction of the blast furnace and the production of crucible steel. This was followed by the open-hearth furnace and then the Bessemer process in England in the mid-19th century.

The modern steelmaking process started with the acid Bessemer method invented in 1856, which solved the problem of mass production of liquid steel for the first time and laid the foundation of modern steelmaking process methods. As a result of the direct interaction between air and molten iron, Bessemer's method of steelmaking had a high smelting speed and became the dominant steelmaking method at that time.

With the invention of the Bessemer process, a new era of mass-produced steel began. Mild steel replaced wrought iron. In 1879, an Englishman, S. G. Thomas, invented the basic air bottom blown converter steelmaking process, which solved the problem of smelting the high-phosphorus pig iron successfully. As many iron ore mines in Western Europe are highly phosphorous, the Thomas process was still used in some steel plants in France, Luxembourg, Belgium and other countries until the late 1970s.

Further refinements in the process, such as basic oxygen steelmaking (BOS), largely replaced earlier methods by further lowering the cost of production and increasing the quality of the final product. Today, steel is one of the most common manmade materials in the world, with more than 1.6 billion tons produced annually. Modern steel is generally identified by various grades defined by assorted standards organizations.

The carbon content of steel is between 0.002% and 2.14% by weight for plain iron-carbon alloys. These values vary depending on alloying elements such as manganese, chromium, nickel, and tungsten. In contrast, cast iron does undergo an eutectic reaction. Too little carbon content leaves (pure) iron quite soft, ductile, and weak. Carbon contents higher than those of steel make a brittle alloy commonly called pig iron. While iron alloyed with carbon is called carbon steel, alloy steel is steel to which other alloying elements have been intentionally added to modify the characteristics of steel. Common alloying elements include manganese, nickel, chromium, molybdenum, boron, titanium, vanadium, tungsten, cobalt, and niobium. Additional elements, most frequently considered undesirable, are also important in steel: phosphorus, sulfur, silicon, and traces of oxygen, nitrogen, and copper.

Plain carbon-iron alloys with a higher than 2.1% carbon content are known as cast iron. With modern steelmaking techniques such as powder metal forming, it is possible to make very high-carbon (and other alloy material) steels, but such are not common. Cast iron is not malleable even when hot, but it can be formed by casting as it has a lower melting point than steel and good castability properties. Certain compositions of cast iron, while retaining the economies of melting and casting, can be heat treated after casting to make malleable iron or ductile iron objects. Steel is distinguishable from wrought iron (now largely obsolete), which may contain a small amount of carbon but large amounts of slag.

Iron is commonly found in the Earth's crust in the form of an ore, usually an iron oxide, such as magnetite and hematite. Iron is extracted from iron ore by removing the oxygen through its combination with a preferred chemical partner such as carbon which is then lost to the atmosphere as carbon dioxide. This process, known as smelting, was first applied to metals with lower melting points, such as tin, which melts at about 250 ℃ (482 ℉), and copper, which melts at about 1,100 ℃ (2,012 ℉), and the combination, bronze, which has a melting point lower than 1,083 ℃ (1,981.4 ℉). In comparison, cast iron melts at about 1,375 ℃ (2,507 ℉). Small quantities of iron were smelted in ancient times, in the solid state, by heating the ore in a charcoal fire and then welding the clumps together with a hammer and in the process squeezing out the impurities. With care, the carbon content could be controlled by moving it around in the fire. Unlike copper and tin, liquid or solid iron dissolves carbon quite readily.

All of these temperatures could be reached with ancient methods used since the Bronze Age. Since the oxidation rate of iron increases rapidly beyond 800 ℃ (1,472 ℉), it is important that the smelting takes place in a low-oxygen environment. The smelting, using carbon to reduce iron oxides, results in an alloy (pig iron) that retains too much carbon to be called steel. The excess carbon and other impurities are removed in a subsequent step.

Other materials are often added to the iron/carbon mixture to produce steel with desired properties. Nickel and manganese in steel add its tensile strength and make the austenite form of the iron-carbon solution more stable, chromium increases hardness and melting temperature, and vanadium also increases hardness while making it less prone to metal fatigue.

To inhibit corrosion, at least 11% chromium is added to steel so that a hard oxide forms on the metal surface; this is known as stainless steel. Tungsten slows the formation of cementite, keeping carbon in the iron matrix and allowing martensite to preferentially form at slower quench rates, resulting in high speed steel. On the other hand, sulfur, nitrogen, and phosphorus are considered contaminants that make steel more brittle and are removed from the steel melt during processing.

Even in a narrow range of concentrations of mixtures of carbon and iron that make a steel, a number of different metallurgical structures, with very different properties can form. Understanding such properties is essential to making quality steel. At room temperature, the most stable form of pure iron is the body-centered cubic (BCC) structure called alpha iron or α-iron. It is a fairly soft metal that can dissolve only a small concentration of carbon. The inclusion of carbon in alpha iron is called ferrite. At 910 ℃, pure iron transforms into a face-centered cubic (FCC) structure, called gamma iron or γ-iron. The inclusion of carbon in gamma iron is called austenite. The more open FCC structure of austenite can dissolve considerably more carbon, as much as 2.1% (38 times that of ferrite) carbon at 1,148 ℃ (2,098.4 ℉), which reflects the upper carbon content of steel, beyond which is cast iron. When carbon moves out of solution with iron, it forms a very hard, but brittle material called cementite (Fe_3C).

When steels with exactly 0.8% carbon (known as a eutectoid steel), are cooled, the austenitic phase (FCC) of the mixture attempts to revert to the ferrite phase (BCC). The carbon no longer fits within the FCC austenite structure, resulting in an excess of carbon. One way for carbon to leave the austenite is to precipitate out of solution as cementite, leaving behind a surrounding phase of BCC iron called ferrite with a small percentage of carbon in solution. The two, ferrite and cementite, precipitate simultaneously producing a layered structure called pearlite, named for its resemblance to mother of pearl. In a hypereutectoid composition (greater than 0.8% carbon), the carbon will first precipitate out as large inclusions of cementite at the austenite grain boundaries until the percentage of carbon in the grains has decreased to the eutectoid composition (0.8% carbon), at which point the pearlite structure forms. For steels that have less than 0.8% carbon (hypoeutectoid), ferrite will first form within the grains until the remaining composition rises to 0.8% of carbon, at which point the pearlite structure will form. No large inclusions of cementite will form at the boundaries in hypoeuctoid steel. The above assumes that the cooling process is very slow, allowing enough time for the carbon to migrate.

As the rate of cooling is increased, the carbon will have less time to migrate to form carbide at the grain boundaries but will have increasingly large amounts of pearlite of a finer and finer structure within the grains; hence the carbide is more widely dispersed and acts to prevent slip of defects within those grains, resulting in hardening of steel. At the very high cooling rates produced by quenching, the carbon has no time to migrate but is locked within the face-centered austenite and forms martensite. Martensite is a highly strained, stressed and supersaturated form of carbon and iron and is exceedingly hard but brittle. Depending on the carbon content, the martensitic phase takes different forms. Below 0.2% carbon, it takes on a ferrite BCC crystal form, but at higher carbon content it takes a body-centered tetragonal (BCT) structure. There is no thermal activation energy for the transformation from austenite to martensite. Moreover, there is no compositional change so the atoms generally retain their same neighbors.

Martensite has a lower density (it expands during the cooling) than austenite, so that the transformation between them results in a change of volume. In this case, expansion occurs. Internal stresses from this expansion generally take the form of compression on the crystals of martensite and tension on the remaining ferrite, with a fair amount of shear on both constituents. If quenching is done improperly, the internal stresses can cause a part to shatter as it cools. At the very least, they cause internal work hardening and other microscopic imperfections. It is common for quench cracks to form when steel is water quenched, although they may not always be visible.

Terms

1. crystalline 晶状的
2. allotropic 同素异形体的
3. body-centered cubic 体心立方
4. face-centered cubic 面心立方
5. pure iron 纯铁
6. hardening agent 硬化剂
7. bloomery furnace 炼铁炉
8. mild steel 低碳钢
9. plain iron-carbon alloy 普通铁碳合金
10. molybdenum 钼
11. boron 硼
12. niobium 铌
13. hematite 赤铁矿
14. smelt 熔炼
15. tin 锡
16. charcoal fire 炭火
17. hardness 硬度
18. ferrite 铁酸盐
19. hypereutectoid 过共析的
20. Bessemer process 贝塞麦炼钢法
21. basic oxygen steelmaking (BOS) 碱性氧气炼钢
22. eutectic reaction 共晶反应(指在一定的温度下,一定成分的液体同时结晶出两种一定成分的固相的反应)

Exercises

1. Why is steel the most useful material?
2. What's the difference between steel and iron?
3. What's the relation between alloy and steel?

Passage B Heat Treatment

There are many types of heat-treating processes available to the steel. The most common are annealing, quenching, and tempering. Heat treatment is effective on compositions above the eutectoid composition (hypereutectoid) of 0.8% carbon. Hypoeutectoid steel does not benefit from heat treatment.

Annealing is the process of heating the steel to a sufficiently high temperature to relieve local internal stresses. It does not create a general softening of the product but only locally relieves strains and stresses locked up within the material. Annealing goes through three phases: recovery, recrystallization, and grain growth. The temperature required to anneal a particular steel depends on the type of annealing to be achieved and the alloying constituents.

Quenching involves heating the steel to create the austenite phase then quenching it in water or oil. This rapid cooling results in a hard but brittle martensitic structure. The steel is then tempered, which is just a specialized type of annealing, to reduce brittleness. In this application, the annealing (tempering) process transforms some of the martensite into cementite, or spheroidite, and hence it reduces the internal stresses and defects. The result is a more ductile and fracture-resistant steel.

When iron is smelted from its ore, it contains more carbon. To become steel, it must be reprocessed to reduce the carbon to the correct amount, at which point other elements can be added. In the past, steel facilities would cast the raw steel product into ingots which would be stored until use in further refinement processes that resulted in the finished product. In modern facilities, the initial product is close to the final composition and is continuously cast into long slabs, cut and shaped into bars and extrusions, and heat-treated to produce a final product. Today, approximately 96% of steel is continuously cast, while only 4% is produced as ingots.

The ingots are then heated in a soaking pit and hot rolled into slabs, billets, or blooms. Slabs are hot or cold rolled into sheet metal or plates. Billets are hot or cold rolled into bars, rods, and wire. Blooms are hot or cold rolled into structural steel, such as I-beams and rails. In modern steel mills, these processes often occur in one assembly line, with ore coming in and finished steel products coming out. Sometimes after a steel's final rolling, it is heat-treated for strength; however, this is relatively rare.

Terms

1. spheroidite 球状渗碳体
2. ingot 铸模

Exercises

1. What is the heat process for?
2. What is the main heat process?
3. Why do we need to heat or cold steel?

Passage C Types of Steel

Carbon steel

Carbon content of carbon steel is about 0. 05% - 0. 70% , as high as 0. 90% in some cases. It can be divided into ordinary carbon structural steel and high-quality carbon structural steel. The former contains more impurities with low prices, and used for normal performance requirements. Most of its carbon content below 0.30% , and the manganese content does not exceed 0. 80% . It is featured by low strength, good plastic and ductility as well as cold deformation performance. Except for a few cases, it is generally not used in heat treatment directly, like bar steel, shaped steel, steel plate and so on. It is mainly used in railway, bridge and all kinds of construction engineering, manufacturing all kinds of metal components, unimportant mechanical parts, and general welding parts that do not need heat treatment. High-quality carbon structural steel has pure steel quality, few impurities, and good mechanical properties, and can be used after heat treatment. According to the content of manganese, they are divided into two groups: ordinary manganese content (less than 0. 80%) and high manganese content (0. 80% - 1. 20%). Carbon content below 0.25% , would be used without heat treatment, or through carburizing, and other treatment, manufacturing small- and medium-sized gear, shaft, and piston pin. The carbon content is 0. 25% - 0. 60% , and the typical steel number is 40,45,40 Mn, 45Mn, etc., after the tempering treatment, the manufacture of various mechanical parts and fasteners. More than 0. 60% carbon content, such as 65,70,85,65Mn, 70Mn, is mostly used as spring steel.

Alloy steel

Alloys are metallic materials with metallic properties that are formed by alloying processes (smelting, mechanical alloying, sintering, vapor deposition, etc.) in which two or more metallic elements or other nonmetallic elements are added based on the metal. But alloys may contain only one metal element, such as steel. (Steel is a general term for ferroalloys with a carbon content between 0.02% and 2.00% in mass.) It should be noted here that alloys are not mixtures in the general concept and can even be pure substances, such as single-phase metal compound alloys. The alloy elements can form solid solutions and compounds, and produce endothermic or exothermic reactions to change the properties of the metal matrix.

The formation of alloys often improves the elemental properties. For example, steel is stronger than its principal component, iron. The physical properties of the alloy, such as density, reactivity, electrical, and thermal conductivity, may be similar to the constituent elements of the alloy, but the tensile and shear strength of the alloy are usually quite different from the properties of the constituent elements. This is due to the very different arrangement of atoms in the alloy and the elemental matter. A small amount of a certain element may have a great effect on the properties of the alloy. For example, impurities in ferromagnetic alloys can change the properties of the alloys.

Unlike pure metals, most alloys have no fixed melting point and the mixture exists in a solid-liquid coexistence state when the temperature is within the melting temperature range. Therefore, it can be said that the melting point of alloys is lower than that of component metals. Among common alloys, brass is an alloy of copper and zinc; bronze is an alloy of tin and copper, used for statues, ornaments and church bells. Some countries use alloys (such as nickel) for their currency. Stainless steels contain a minimum of 11% chromium, often combined with nickel, to resist corrosion. Some stainless steels, such as the ferritic stainless steels are magnetic, while others, such as the austenitic, are nonmagnetic. Corrosion-resistant steels are abbreviated as CRES.

Some more modern steels include tool steels, which are alloyed with large amounts of tungsten and cobalt or other elements to maximize solution hardening. It allows the use of precipitation hardening and improves the alloy's temperature resistance. Tool steel is generally used in axes, drills, and other devices that need a sharp, long-lasting cutting edge. Other special-purpose alloys include weathering steels such as Corten, which weather by acquiring a stable, rusted surface, and so can be used unpainted. Maraging steel is alloyed with nickel and other elements, unlike most steel that contains little carbon (0.01%). This creates a very strong but still malleable steel.

Eglin steel uses a combination of over a dozen of different elements in varying amounts to create a relatively low-cost steel for use in bunker buster weapons.

Hadfield steel (after Sir Robert Hadfield) or manganese steel contains 12% - 14% manganese which when abraded strain-hardens to form a very hard skin which resists wearing. Examples include tank tracks, bulldozer blade edges, and cutting blades on the jaws of life.

Terms

1. spring steel 弹簧钢
2. endothermic 吸热的
3. exothermic 放热的
4. maraging steel 马氏体时效钢
5. Corten 考登钢
6. eglin steel 埃格林钢

Exercises

1. What are advantages of alloy?
2. What are characteristics of stainless steel?
3. What can alloy do?

Passage D Steel History

Steel was known in antiquity and was produced in bloomeries and crucibles. The earliest known production of steel is seen in pieces of ironware excavated from an archaeological site in Anatolia (Kaman-Kalehöyük) and are nearly 4,000 years old, dating from 1,800 BC. Metal production sites in Sri Lanka employed wind furnaces driven by the monsoon winds, capable of producing high-carbon steel. Large-scale Wootz steel production in Tamilakam using crucibles and carbon sources as the plant Avāram occurred by the 6th century BC, the pioneering precursor to modern steel production and metallurgy. With time going by, not until Bessemer had invented the new steelmaking that the steel has been manufacturing today. Because of its better physical, chemical and mechanical properties than the original pig iron, steel was soon widely used. However, due to the limitations of technical conditions, the application of steel was limited by the output of steel. Until the 18th century after the Industrial Revolution, the application of steel made rapid progress.

Wuhan Iron and Steel Group Corporation (Wisco) is the first super-large steel joint enterprise established after the founding of the People's Republic of China. It began construction in 1955 and completely put into operation on September 13, 1958. It is an important state-owned backbone enterprise directly managed by the Central Government and the State-owned Assets Supervision and Administration Commission of the State Council. The plant is located in the eastern suburb of Wuhan City, Hubei Province, on the south bank of the Yangtze River, covering an area of 21.17 square kilometers. Wisco owns a complete set of advanced steel production technology and equipment, such as mining, coking, iron making, steelmaking, steel rolling, logistics and supporting public facilities. After the joint reorganization of Esteel, Liusteel and Kunming Steel, wisco has become a large enterprise group with a production scale of nearly 40 million tons, ranking the fourth place in the world's steel industry.

Wisco has three main industries, namely steel manufacturing industry, high and new technology industry and international trade. Steel products are mainly about hot rolled coil, hot rolled section steel.

There are hundreds of varieties of hot rolled heavy rail, medium-thick plate, cold rolled coil plate, galvanized plate, tin plate, cold rolled and non-oriented silicon steel sheet, colored coated steel sheet, high-speed wire rod and so on. Among them, the cold rolled silicon steel sheet and ship plate steel was awarded as "China famous brand product", car plate, bridge steel, pipeline steel, pressure vessel steel, container steel, cord steel, refractory weathering steel, electrical steel and other high-quality brand-name products enjoy extensive reputation in domestic and foreign market, "wisco brand was named the national famous trademark". Wisco has been awarded National Technological Innovation Award, National Quality Management Award, National Advanced Enterprise of Quality and Efficiency, National Advanced Unit of Customer Satisfaction, National Outstanding Contribution Award of Enterprise Management, National Civilized Unit and one of the top ten typical central enterprises.

Exercises

1. Name two or three big events in the steel history.
2. Why did Wisco come to decline and reorganize later?
3. How can the iron and steel corporation develop better?

Passage E Reading Material: Reduction of Iron Oxides

In view of its practical importance to the understanding and control of iron-making processes, a great deal of research has been done on the gaseous reduction of iron oxides and iron ores. Because of the porous nature of iron oxides and the reduction products, the interpretation of the reduction rate data is inherently complex.

The formation of product layers during the gaseous reduction of dense sintered hematite and magnetite pellets or natural dense iron ore particles is a well-known phenomenon.

In several studies made in the early 1960s, it was found that the thickness of the reduced iron layer, encasing the iron oxide core of the pellet, increased linearly with the reduction time. The measured rates were interpreted in terms of the rate-controlling chemical reaction at the iron wustite interface, the diffusive fluxes of gases through the porous layers were assumed to be relatively fast. On the other hand, Warner and Spitzer et al. have expressed the view that the rate of gaseous reduction is much affected by the gaseous diffusional process, e. g. the gas-film resistance at the pellet surface and particularly the resistance to diffusion in the porous product layers. The rate measurements made by Turkdogan and Vinters on the reduction of hematite pellets in H_2-H_2O and CO-CO_2 mixtures have clearly demonstrated that the rate-controlling effects of gas diffusion into the pores of the oxide granules or pellets and through the porous iron layers dominate reaction kinetics.

3. Nonferrous Metals

Passage A An Overview of Nonferrous Metals

In metallurgy, a nonferrous metal is a metal, including alloys, which does not contain iron (ferrite) in appreciable amounts.

Nonferrous metals are generally more costly than ferrous metals, and are used because of desirable properties such as low weight (e.g. aluminum), higher conductivity (e.g. copper), non-magnetic property or resistance to corrosion (e.g. zinc). Some nonferrous materials are also used in the iron and steel industries. For example, bauxite is used as flux for blast furnaces, while others such as wolframite, pyrolusite, and chromite are used in making ferrous alloys.

Important nonferrous metals include aluminum, copper, lead, nickel, tin, titanium and zinc, and alloys such as brass. Precious metals such as gold, silver and platinum, and exotic or rare metals such as cobalt, mercury, tungsten, beryllium, bismuth, cerium, cadmium, niobium, indium, gallium, germanium, lithium, selenium, tantalum, tellurium, vanadium, and zirconium are also nonferrous. They are usually obtained through minerals such as sulfides, carbonates, and silicates. Nonferrous metals are usually refined through electrolysis.

Recycling and pollution control

Due to their extensive use, nonferrous scrap metals are usually recycled. The secondary materials in scrap are vital to the metallurgy industry, as the production of new metals often needs them. Some recycling facilities resmelt and recast nonferrous materials; the dross is collected and stored onsite while the metal fumes are filtered and collected. Nonferrous scrap metals are sourced from industrial scrap materials, particle emissions, and obsolete technology (for example, copper cables) scrap.

Ancient history

Nonferrous metals were the first metals used by humans for metallurgy. Gold, silver and copper existed in their native crystalline yet metallic form. These metals, though rare, could be found in quantities sufficient to attract the attention of humans. Less susceptible to oxygen than most other metals, they can be found even in weathered outcroppings. Copper was the first metal to be forged; it was soft enough to be fashioned into various objects by cold forging and could be melted in a crucible. Gold, silver, and copper replaced some of the functions of other resources, such as wood and stone, because they can be shaped into various forms for different uses. Due to their rarity, these gold, silver and copper artifacts were treated as luxury items and handled with great care. The use of copper also heralded the transition from the Stone Age to the Copper Age. The Bronze Age, which succeeded the Copper Age, was again heralded by the invention of bronze, an alloy of copper with the nonferrous metal tin.

Mechanical and structural use

It is used in residential, commercial, and industrial industry. Material selection for a mechanical or structural application requires some important considerations, including how easily the material can be shaped into a finished part and how its properties can be either intentionally or inadvertently altered in the process. Depending on the end use, metals can be simply cast into the finished part, or an intermediate form, such as an ingot, and then worked, or wrought, by rolling, forging, extruding, or other deformation process. Although the same operations are used with ferrous as well as nonferrous metals and alloys, the reaction of nonferrous metals to these forming processes is often more severe. Consequently, properties may differ considerably between the cast and wrought forms of the same metal or alloy.

Terms

1. zinc 锌	2. bauxite 铝矾土,铝土矿
3. wolframite 黑钨矿,钨锰铁矿	4. pyrolusite 软锰矿
5. lead 铅	6. brass 黄铜
7. platinum 铂	8. mercury 汞
9. bismuth 铋	10. cerium 铈
11. indium 铟	12. gallium 镓
13. germanium 锗	14. lithium 锂
15. selenium 硒	16. tantalum 钽
17. tellurium 碲	18. sulfide 硫化物
19. carbonate 碳酸盐	20. silicate 硅酸盐
21. electrolysis 电解	22. forge 锻造,熔炉
23. crucible 坩埚	24. bronze 青铜

Exercises

1. What's the difference between ferrous metals and nonferrous metals?
2. Please summarize the reason of the recycling of nonferrous metals.
3. What's the significance of the use of copper?

Passage B Classification of Nonferrous Metals

For the convenience of research, nonferrous metals are classified into five categories according to their properties, ore characteristics, uses, and the time of use.

Light nonferrous metals

Light nonferrous metals include aluminum (Al), magnesium (Mg), sodium (Na), potassium (K), calcium (Ca), strontium (Sr), and barium (Ba). The common characteristics of these metals are: low density (0.53 - 4.5), high chemical activity, and stable compounds with oxygen, sulfur, carbon and halogen.

Heavy nonferrous metals

Heavy nonferrous metals include copper (Cu), lead (Pb), zinc (Zn), nickel (Ni), cobalt (Co), tin (Sn), antimony (Sb), mercury (Hg), cadmium (Cd), and bismuth (Bi). They are characterized by high density, such as 11.34 for lead. According to their characteristics, each heavy nonferrous metal has its special application range and purpose in various sectors of the national economy. For example, copper is the basic material for the military industry and electrical equipment, and nickel and cobalt are important alloying elements for making high-temperature alloys and stainless steels.

Precious metals

Gold (Au), silver (Ag) and platinum group elements [platinum (Pt), rhodium (Rh), palladium (Pd), osmium (Os), iridium (lr), ruthenium (Ru)] are known as precious metals. These metals are more expensive than ordinary metals because of their stability to oxygen and other reagents, and low content in the crust, which makes mining and extracting them difficult. Except for gold, silver and platinum, which have their own minerals and can be produced partly from ore, most of these metals are recovered from the by-products (anode slime) of smelteries such as copper, lead, zinc, and nickel. They are characterized by high density (10.4 - 22.4) in which platinum, iridium, and osmium are the heaviest metallic elements. Besides, they have high melting point (916 ℃ - 3,000 ℃), stable chemical properties, and can resist acid and alkali corrosion. In addition, gold and silver have a high degree of forgeability and plasticity, palladium and platinum also have good plasticity, and the others are brittle metals. Gold and silver have good electrical and thermal conductivity, but platinum group elements have very low electrical and thermal conductivity. Precious metals are widely used in electrical and electronic industries, space technology, high-temperature instruments and contact agents.

Semi metals

Semi metals generally refer to the five elements of silicon (Si), selenium (Se), tellurium (Te), arsenic (As), and boron (B), which are classified as nonferrous metals in the industry. The physical and chemical properties of these metals are between metals and nonmetals. For example, arsenic is a nonmetal, but it can conduct heat and electricity. These metals have different uses according to their characteristics. For example, silicon is one of the main materials of semiconductors, arsenic, and tellurium; selenium is the raw material for manufacturing compound semiconductors; boron is an important alloying element.

Rare metals

Rare metals are characterized by late discovery, difficulty in extraction, and late application in the industry. Due to the large number, to facilitate research, they are divided into five categories according to their nature, extraction methods, and characteristics existing in the crust.

- **Light rare metals**

Light rare metals [including five metals: lithium (Li), beryllium (Be), rubidium (Rb), cesium (Cs), and titanium (Ti)] are characterized by low density, such as lithium at 0. 534 and high chemical activity. The oxides and chlorides of these metals have high chemical stability and are difficult to be reduced. They are usually produced by molten salt electrolysis.

- **High melting rare metals**

High melting rare metals (also known as rare refractory metals), including eight metals: zirconium (Zr), hafnium (Hf), vanadium (V), niobium (Nb), tantalum (Ta), tungsten (W), molybdenum (Mo), and rhenium (Re). They are characterized by high melting point, such as the melting point of tungsten is 3,410 ℃; They have high hardness and strong corrosion resistance and can form very hard and refractory stable compounds with some nonmetals, such as carbide, nitride, silicide and boride. These compounds are important materials for the production of cemented carbide.

- **Rare-dispersed metals**

Rare-dispersed metals (also known as scattered metals), including gallium

(Ca), indium (In), thallium (Tl), and germanium (Ce). Their characteristic is that this kind of metal is very scattered in the Earth's crust and is often associated with other mineral deposits. But its output is very small and has no industrial value. It is usually extracted from the waste of metallurgical plants or chemical plants, such as anode slime of electrolytic copper, slag and soot from smelting lead, zinc and aluminum.

- **Rare earth metals**

There are 17 rare earth metals, including scandium (Sc), yttrium (Y), lanthanum (La), cerium (Ce), praseodymium (Pr), neodymium (Nd), promethium (Pm), samarium (Sm), europium (Eu), gadolinium (Gd), terbium (Tb), dysprosium (Dy), holmium (Ho), erbium (Er), thulium (Tm), ytterbium (Yb), and lutetium (Lu). Among them, what from lanthanum to europium are light rare earths, and what from gadolinium to lutetium plus scandium and yttrium are heavy rare earths. In the 18th century, only rare earth oxides that looked like alkaline earth (such as calcium oxide) could be obtained. Therefore, it was named "rare earth" and it is still used today. These metals have the same atomic structure, so their chemical properties are very similar. They are always associated with each other in the ore. During the extraction process, they need to go through complicated operations to extract and separate them one by one.

- **Rare radioactive metals**

As for radioactive rare metals, there are 6 kinds of natural radioactive elements: polonium (Po), radium (Ra), actinium (Ac), thorium (Th), protactinium (Pa), and uranium (U), and 13 kinds of artificial trans-uranium elements: francium (Pr), technetium (Tc), neptunium (Np), plutonium (Pu), americium (Am), curium (Cm), berkelium (Bk), californium (Cf), einsteinium (Es), fermium (Fm), mendelevium (Md), nobelium (No) and lawrencium (Lw). Natural radioactive elements often coexist in ore, and they are often associated with rare earth metal ore. This type of metal plays an extremely important role in the atomic energy industry.

Terms

1. light nonferrous metal 轻有色金属
2. sodium 钠
3. potassium 钾
4. calcium 钙
5. strontium 锶
6. barium 钡
7. halogen 卤素
8. heavy nonferrous metal 重有色金属
9. antimony 锑
10. precious metal 贵金属
11. rhodium 铑
12. palladium 钯
13. iridium 铱
14. ruthenium 钌
15. anode slime 阳极泥
16. smeltery 冶炼厂
17. acid 酸
18. alkali 碱
19. corrosion 腐蚀
20. forgeability 可锻性
21. plasticity 可塑性
22. brittle metal 脆性金属
23. semimetal 半金属
24. silicon 硅
25. arsenic 砷
26. rare metal 稀有金属
27. light rare metal 轻稀有金属
28. rubidium 铷
29. cesium 铯
30. oxide 氧化物
31. chloride 氯化物
32. molten salt electrolysis 熔盐电解法
33. high melting rare metal 高熔点稀有金属
34. rare refractory metal 难熔稀有金属
35. carbide 碳化物
36. nitride 氮化物
37. silicide 硅化物
38. boride 硼化物
39. cemented carbide 硬质合金
40. rare-dispersed metal 稀有分散金属
41. scattered metal 分散金属
42. mineral deposit 矿床
43. metallurgical plant 冶金工厂
44. chemical plant 化工厂
45. soot 煤烟
46. rare earth metal 稀土金属
47. scandium 钪
48. yttrium 钇
49. lanthanum 镧
50. praseodymium 镨
51. neodymium 钕
52. promethium 钷
53. samarium 钐
54. europium 铕
55. gadolinium 钆
56. terbium 铽
57. dysprosium 镝
58. holmium 钬

59. erbium 铒
60. thulium 铥
61. ytterbium 镱
62. lutetium 镥
63. calcium oxide 氧化钙
64. rare radioactive metal 稀有放射金属
65. polonium 钋
66. actinium 锕
67. thorium 钍
68. protactinium 镤
69. trans-uranium 超铀的
70. francium 钫
71. technetium 锝
72. neptunium 镎
73. plutonium 钚
74. americium 镅
75. curium 锔
76. berkelium 锫
77. californium 锎
78. einsteinium 锿
79. fermium 镄
80. mendelevium 钔
81. nobelium 锘
82. lawrencium 铹

Exercises

1. What's the total number of nonferrous metals in Passage B?

2. Please illustrate the similarities and differences between precious metals and rare metals.

3. What are brittle metals included?

4. Please introduce the application about a kind of nonferrous metals or alloys that you are familiar with in your daily life.

Passage C Nonferrous Alloys

The nonferrous metal is used as the matrix, and one or more other metals or nonmetallic elements are added to form a substance that has the generality of the base metal as well as certain specific properties, which is called a nonferrous alloy. Besides, it also can be seen as a mixture of two or more metallic elements that does not contain iron.

Classification of nonferrous alloys

There are many classification methods for nonferrous alloys, and the following two are the general methods.

- **In terms of the system of nonferrous alloys**

It can be classified as heavy nonferrous alloys, including copper alloys such as red copper, brass, cupronickel, nickel alloys, zinc alloys, lead alloys and tin alloys; light nonferrous alloys, including aluminum alloys, magnesium alloys, etc; noble metal alloys, including silver alloys and platinum group alloys; rare metal alloys, including titanium alloys, tungsten alloys, molybdenum alloys, niobium alloys and rhenium alloys.

- **In terms of the application of nonferrous alloys**

It can be classified as deforming alloys (alloys for pressure processing), casting alloy, bearing alloy, printing alloy, cemented carbide, solder alloy, and master alloy.

Property of nonferrous alloys

A distinguishing feature of nonferrous alloys is that they are highly malleable (i. e., they can be pressed or hammered into thin sheets without breaking). Nonferrous simply implies a metal other than iron or steel.

Nonferrous alloys have one valuable advantage over ferrous alloys and metals, which is that they are highly corrosion- and rust-resistant because they do not have any iron content in them. Consequently, these materials are suitable for highly corrosive environments such as liquid, chemical, and sewage pipelines. Nonferrous alloys and metals are also nonmagnetic, making them suitable for many electrical and electronic applications. Some commonly used nonferrous metals and alloys are copper, zinc, aluminum, lead, nickel, cobalt, chromium, gold, and silver.

Terms

1. cupronickel 铜镍合金
2. deforming alloy 变形合金
3. casting alloy 铸造合金
4. bearing alloy 轴承合金
5. printing alloy 印刷合金
6. solder alloy 焊料合金
7. master alloy 中间合金

4. Metal Extraction Processes

Passage A

Part One Introduction of Metal Extraction Processes

Metals occur in the Earth's crust in the following chemical states:

(1) Oxides, e. g. Fe_2O_3, TiO_2, Cu_2O, SnO_2;

(2) Sulphides, e. g. PbS, ZnS, Cu_2S, Ni_3S_2, HgS;

(3) Oxysalts, such as silicates, sulphates, titanates, carbonates, e. g. Fe_2CO_3, $ZrSiO_3$; and to a lesser extent in other forms such as in "native" (elemental) form or as arsenides, e. g. $PtAs_2$.

By far the most important groups with respect to quantity and occurrence are sulphides andoxides. Apart from the precious metals (Au, Ag, Pt), metals rarely occur in the uncombined or "native" form. This is due to the reactive nature of metals which combine with the environment producing such compounds as oxides and sulphides. Precious metals are least reactive.

Metal ores are concentrations of the above metal compounds associated with other unwanted minerals (gangue) such as silicates. It is therefore necessary to separate the metal-bearing component (value) from the unwanted gangue prior to the extraction process. If this is not done, the subsequent extraction of the metal will be less efficient and more costly, and it will be more difficult to produce the metal in a state of high purity.

Once the metal-beating constituent and the gangue have been separated, the concentrate produced can be subjected to the extraction process. There are three main routes that can be used:

(1) Pyrometallurgy—incorporates smelting, converting and fire refining of the metal concentrate;

(2) Hydrometallurgy—provides the metal in the form of an aqueous solution followed by subsequent precipitation of the metal;

(3) Electrometallurgy—uses electrolysis to extract the metal. Electrowinning is

the extraction of the metal from the electrolyte while electrorefining is the refining of the impure metal which is in the form of the anode.

The choice of extraction route will largely depend on the cost per tonne of metal extracted which, in turn, will depend on the type of ore, availability and cost of fuel (coal, oil, natural gas or electricity), production quantity and rate and the required metal purity. Unless cheap hydroelectric power is available or the metal is highly reactive, such as aluminum, electrometallurgy is an expensive route but it usually provides the metal in an extremely pure state of about 99.9% +. Electro-refining is often used as a final refining process after pyrometallurgical extraction. Hydrometallurgy tends to be slower than pyrometallurgical extraction process and reagent costs tend to be high but it is ideal for extracting metals from lean ores. Again, a final electro-refining process is often adopted. Owing to the abundance and relatively low cost of fossil fuels (oil, coal, coke and natural gas) and the fact that pyrometallurgy is more adaptable than hydrometallurgy and electrometallurgy to high production rates, pyrometallurgical processes provide the main routes for the extraction of metals.

Part Two　Pyrometallurgical Extraction Processes

1. Extraction of copper from copper sulphide concentrates

Ore preparation and flotation normally provide a Cu_2S concentrate containing 25% Cu or more from an ore which "as mined" may have contained as little as 0.5% to 2% Cu. The main impurity is iron and is often associated with nickel, gold and silver, and traces of Zn, Se, As, Sb, Te, Co, Sn, and Pb.

Partial roasting

Partial roasting of the concentrate is conducted in multi-hearth, flash, or fluidised bed units. The main aim of the partial roasting process is to reduce the sulphur level but ensuring all of the copper is present as Cu_2S and part of the iron are retained as FeS. The reason for retaining some FeS in the calcine is to provide autogenous processing due to the large exothermic reaction between FeS andoxygen during the converting stage. Some preferential oxidisation of certain impurity metal sulphides is also encouraged in partial roasting, e. g. :

$$2ZnS + 3O_2 \rightarrow 2ZnO + 2SO_2$$

These oxides will be collected in the slag on the subsequent smelting.

Smelting

The calcine product is smelted with an acid (silica) slag in a reverberatory or flash smelting unit. A sulphide matte is produced containing Cu_2S-FeS plus precious metals (Ag, Au) and other impurity sulphides which were not oxidised during partial roasting. If arsenic and antimony are present, a liquid speiss layer is formed. The basic FeO produced in the partial roasting operation reacts with the acid slag, $FeO + SiO_2 \rightarrow FeSiO_3$, which also collects other impurities such as amphoteric oxides (ZnO).

Converting

The sulphide matte is transferred to a converter through which is blown air and/or oxygen. The impurity metal sulphides are preferentially oxidised, and the oxidation of the remaining FeS which was deliberately retained during partial roasting provides sufficient heat to make the converting process autogenous.

This reaction is only possible because the mutual reduction reaction has a negative ΔG value at 1,200 ℃. The copper content after converting is approximately 98% and is called blister copper due to the bubbling of sulphur dioxide through the melt producing a blister effect. The precious metals, as would be expected from their free energies of oxidation, are not oxidised and are collected in the blister copper.

Fire refining

The blister copper containing the precious metals and small amounts of other impurity elements is subjected to controlled air blowing in a hearth furnace with possible use of oxidizing slags and fluxes. More reactive metals are oxidised and removed to the slag until the copper begins to oxidise. At this stage, any copper oxide formed is reduced by poling the melt with trees trunks or bubbling gaseous NH_3 or gaseous hydrocarbons through the melt. The hydrocarbons released from burning the trees produces carbon monoxide and hydrogen which reduce the copper oxide, returning the copper to the melt from the slag. Gaseous ammonia dissociates with hydrogen and nitrogen, and gaseous hydrocarbons have the same effect, i. e.

There is now a problem of reduction of other oxides which have similar thermodynamic properties to copper and which have separated to the slag as impurity oxides. Therefore, careful alternating of controlled oxidation (air blowing) and reduction (poling) is carried out to provide maximum copper yield with a minimum impurity level. Final deoxidation may be achieved by addition of a phosphorus (15%)-copper alloy, removing the oxygen as P_2O_5. Alternatively, if residual phosphorus is to be avoided due to its present reducing the electrical conductivity of copper, lithium or carbon may be used as the deoxidant.

Fire refining will produce a copper purity of about 99.5%. If a higher purity is required, electrorefining must be used.

2. Extraction of magnesium from magnesium oxide concentrates

Magnesium occurs naturally in the Earth's crust as magnesite ($MgCO_3$), dolomite ($MgCO_3$, $CaCO_3$) and brucite [$Mg(OH)_2$] and in sea water (0.13%) as $MgCl_2$.

Although the major tonnage of magnesium production is by electrolysis of fused magnesium chloride, some thermal reduction of MgO is carried out by using silicon (metallothermic) or carbon (carbothermic) as the reducing agent.

Consultation of the ΔG-T diagram for oxide formation indicates that MgO reduction with silicon is impossible due to the positive ΔG value for the reaction at 1,200℃.

$$2MgO(s) + Si(s) = 2Mg(g) + SiO_2(s) \ (\Delta G^{\ominus}_{1,200\ ℃} = +244\ kJ \cdot mol^{-1})$$

The above reaction assumes that the reactants and products are in their standard thermodynamic states, i. e. unit activity (pure) if solid or liquid and a partial pressure of 1 atm if in gaseous form. By manipulating these activities and partial pressures and applying the van't Hoff isotherm for non-equilibrium conditions, it is possible to achieve a negative ΔG value for the above reaction and therefore provide the reduction of MgO with Si. Applying the equation to the above reaction at 1,200 ℃:

This condition can be readily achieved by lowing the partial pressure of the magnesium vapor produced using vacuum techniques while the activity of SiO_2 is lowered by additions of lime (CaO) to the slag which collects the silica, i. e. formation of a basic slag. In practice the lime is added in the form of calcined dolomite MgO·CaO (Pidgeon process). The silicon is added as high-grade ferrosilicon which is cheaper to produce than pure silicon and also improves fluidity by lowering the melting point of the slag. The reaction is conducted in an externally heated Ni-Cr retort similar to the horizontal retort used in zinc extraction.

$$2MgO \cdot CaO(s) + Fe\text{-}Si(s) \rightarrow 2Mg(g) + Ca_2FeSiO_4$$

Again, several retorts are placed in batches. Since magnesium is the only gaseous product there is no danger of a reversion reaction to MgO during cooling.

By consultation of the $\Delta G^{\ominus}$-T relationships for the carbothermic reduction of MgO the minimum reduction temperature is 1,850 ℃:

$$MgO(s) + C(g) \rightarrow Mg(g) + CO(g)$$

If the magnesium vapor cools below 1,850 ℃ the above reaction will reverse. Therefore, the Mg vapor must be rapidly "shock cooled", with hydrogen or natural gas to prevent considerable reversion to MgO. The magnesium vapor condenses as a fine dust covered with oxide and which easily catches fire on exposure to air (pyrophoric). For this reason, the magnesium dust is refined by vacuum sublimation which makes this process more expensive than the Pidgeon process or the electrolytic process.

Part Three Hydrometallurgical Extraction Processes

"Hydro" means water, and "hydrometallurgy" is therefore the art and science of aqueous methods of extracting metals from their ores. This covers a large variety of processes ranging from the leaching of ores or roasted sulfides through the purification of the solutions to the winning of the metals or their compounds by chemical or electrochemical precipitation. Electrolytic refining of impure metals may sometimes be classified as hydrometallurgy.

Hydrometallurgy is a relatively recent technique, as compared with the ancient pyrometallurgical methods. Modern hydrometallurgy, can be traced back to the end of the 19th century when two major operations were discovered: the cyanidation process for gold and silver and the Bayer Process for bauxite. Later, in the 1940s, a breakthrough came during the Manhattan Project in the USA in connection with uranium extraction. Since then, it has been advancing progressively and even replacing some pyrometallurgical processes.

Sometimes a purification/concentration operation is conducted prior to precipitation. These processes are aiming to obtain a pure and a concentrated solution from which the metal values can be precipitated effectively.

The advantages of hydrometallurgy can be summarized as the following points:

(1) Metals may be obtained directly in a pure form from the leach solution, e. g. by precipitation with hydrogen under pressure, cementation, or electrolysis.

(2) If processes involving amalgam metallurgy are applied, high-purity metals may be recovered from impure leach solutions.

(3) The siliceous gangue in the ore is unaffected by most leaching agents; whereas in pyrometallurgical smelting processes this gangue must be slagged.

(4) Corrosion problems are relatively mild in hydrometallurgy as compared with the deterioration of refractory linings in furnaces, and the necessity for periodic shutdown and replacement.

(5) Most hydrometallurgical processes are carried out at room temperature, and therefore there is no consumption of large amounts of fuel as in pyrometallurgy.

(6) Handling of leaching products is much cheaper and easier than handling molten mattes, slags, and metals.

(7) Hydrometallurgical processing is especially suitable for the treatment of low-grade ores.

(8) A hydrometallurgical process may start on a small scale and expand as

required; however, a pyrometallurgical process usually must be designed as a large-scale operation, since it is more economical to build one big furnace than multiple small ones with the same capacity.

(9) Hydrometallurgical plants usually do not pollute the environment as do smelters. This factor is playing an important role at present due to the strict anti-air-pollution laws.

Some difficulties, however, may be faced when treating an ore by hydrometallurgical methods. Thus, difficulties may arise in separating the insoluble gangue from the leach solution. Also, very small amounts of impurities in the leach solution may badly affect the electrodeposition of a metal, and therefore necessitate a thorough preliminary purification. Further, hydrometallurgical processes are relatively slow, since they are carried out usually at room temperature, while pyrometallurgical operations are fast, because they are carried out at high temperatures.

Part Four Electrometallurgical Extraction Processes

There are at least two processes involved in electrometallurgy. Electrolysis is the use of electric energy to recover or to refine a metal from different media (aqueous, nonaqueous, or fused salt medium), and electric heating is the use of electric energy in smelting processes.

In this text we shall discuss the application of electrolysis to the winning and refining of metals. Electrolytic winning is important for the very reactive light metals: aluminum and magnesium, which almost exclusively are produced by electrolysis of fused salts. For other metals such as copper and zinc, electrolytic winning from aqueous solutions represents an alternative to pyrometallurgical processes. Finally, electrolytic refining, with aqueous electrolytes or fused salts, is important for the production of high-purity copper and aluminum as well as for the recovery of valuable impurities, such as silver and gold from copper.

Ionic transport

An example of an electrolytic cell is shown in Fig. 2-1. Here two copper electrodes are immersed in a solution of $CuSO_4$. At the electrodes the following reactions take place:

$$Cu = Cu^{2+} + 2e^- \text{ (anode)}$$

$$Cu^{2+} + 2e^- = Cu \text{ (cathode)}$$

In the absence of other electrolytic reactions and in the absence of electronic conduction, a total of N (Avogardro's) electrons is transferred for every gram equivalent of copper. Measurements have shown that this corresponds to:

$$F = 96,500 \text{ coulombs} = 26.8 \text{ Ah}$$

For all per gram equivalent of ions transferred. The quantity F is called one Faraday, and is the charge of N (Avogardro's) electrons.

In the electrolyte the dominating ions are Cu^{2+} and SO_4^{2-}. Under the influence of the electric field the positively charged cations move in the direction of the cathode, and the negatively charged anions in the direction of the anode. During the passage of one Faraday, a total number of n_c gram equivalents of cation and n_a gram equivalents of anion will cross a plane perpendicular to the current flow. We call these numbers the cationic and anionic transport numbers.

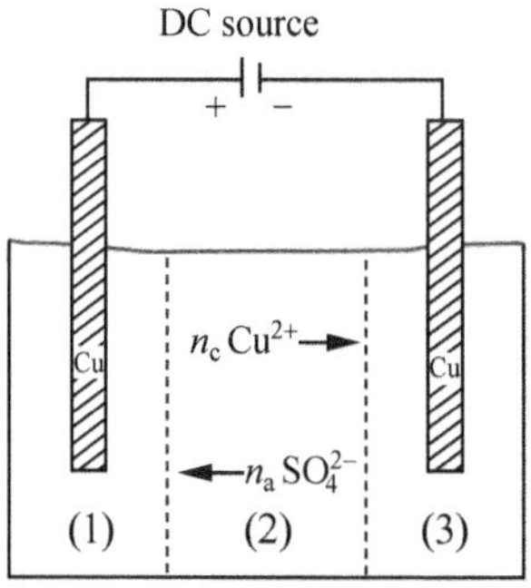

Fig. 2-1 Principle of electrolytic transport:
(1) anolyte; (2) main electrolyte; (3) catholyte

To evaluate these numbers, we imagine the electrolyte divided into three sections: (1) the space around the anode (the anolyte), (2) the main electrolyte, and (3) the space around the cathode (the catholyte). During the flow of one Faraday through the cell, a total of one gram equivalent of copper goes into solution at the anolyte, and one gram equivalent is deposited on the cathode. From the anolyte a total of n_c gram equivalents of copper move into the main electrolyte and the same amount moves into the catholyte. From the catholyte a total of n_a gram equivalents of sulfate ions move into the main electrolyte and further into the anolyte. It is easily seen that, in order to maintain electroneutrality in the anolyte and catholyte, the sum of $n_c + n_a$ must be unity. Thus, the transport numbers give fraction of current which is carried by the two types of ions.

Since, by the passage of one Faraday, the total number of copper equivalents that are added to the anolyte or removed from the catholyte is $1 - n_c$, the transport numbers may be determined from the change in chemical composition of the anolyte or catholyte after the passage of one Faraday.

Notice the high cationic transport number for the acids, which is the result of the high mobility of the hydrogen ion. Likewise, the low value for NaOH shows a correspondingly high value for the OH^- ion mobility. Notice also that for KCl the cationic and anionic transport numbers are closely equal, a fact that is utilized in certain electrochemical measurements.

Electrolytic cells

We may classify electrolytic cells into two main groups:

(1) Refining cells (transference cells)
(2) Production cells (cells without transference)

The example which was shown in Fig. 2-1 illustrated the principle of a refining cell. If the anode and cathode reactions are added, we get the total cell reaction: Cu = Cu, i. e. the reaction has been the transfer of one mole of copper from the anode to the cathode. In a transference cell the two electrodes need not be identical. Very often a metal is transferred from an impure anode or from an alloy or compound to a cathode of the pure metal.

Whereas the former cell would require electric energy to react from left to right, the latter cell may react spontaneously and produce electric energy. It is therefore also called a galvanic cell. In principle any electrolytic cell, including transference cells, may be operated as a galvanic cell by reversing the electric current flow and by supplying the necessary reactants at the electrodes. In cases where all products of electrolysis remain on the electrodes the cell may be operated alternately as an electrolytic cell and as a galvanic cell, i. e. we will have a storage battery.

Terms

1. sulphide 硫化物
2. oxysalt 含氧盐
3. silicate 硅酸盐
4. sulphate 硫酸盐
5. titanate 钛酸盐
6. carbonate 使渗碳,使碳化
7. arsenide 砷化物
8. gangue 脉石
9. pyrometallurgy 高温冶金学
10. hydrometallurgy 湿法冶金学
11. aqueous 水的,含水的
12. precipitation 沉淀,沉析,沉降
13. electrolyte 电解液,电解质
14. anode 阳极,正极
15. reagent 试剂,反应物
16. coke 焦炭,焦煤
17. multi-hearth unit 多膛炉
18. oxidisation 氧化

19. amphoteric 两性的
20. amphoteric oxide 两性氧化物
21. turbulence 湍流
22. oxygen blow 吹氧
23. reaction rate 反应速率
24. activity 活跃度
25. driving force 驱动力
26. mutual reduction reaction 相互还原反应
27. content 含量
28. blister (金属的)砂眼,泡疤
29. blister copper 粗铜
30. sulphur dioxide 二氧化硫
31. reactive metal 活泼金属
32. ammonia 氨
33. dissociate 使分离,使离解
34. deoxidation 脱氧
35. dolomite 白云石
36. brucite 水镁石
37. tonnage 吨
38. reactant 成分;试剂,反应物
39. isotherm 等温线
40. lime 石灰
41. ferrosilicon 硅铁(合金)
42. carbothermic reduction 碳热还原
43. pyrophoric 自燃的
44. sublimation 升华
45. leach 浸出
46. purification 净化;提纯
47. pyrometallurgical 高温冶金的
48. cyanidation 氰化法
49. bauxite 铝矾土; 铝土矿
50. dissolution 溶解
51. concentration 浓度; 浓缩
52. concentrated 浓缩的
53. cementation 烧结;渗碳;置换沉淀
54. amalgam 汞齐,汞合金
55. silicious 含硅的;硅质的
56. slag 使成渣
57. deterioration 磨损,损坏
58. lining 内衬
59. shutdown 关机;停工;关门
60. molten matte 熔锍
61. insoluble 不溶解的
62. electrodeposition 电解沉积
63. necessitate 使成为必需,需要;迫使
64. nonaqueous 非水的
65. fused salt 熔盐
66. electrolytic winning 电解提炼
67. electrolytic refining 电解精炼
68. electrolytic cell 电解槽
69. electrode 电极
70. anolyte 阳极(电解)液
71. catholyte 阴极(电解)液
72. electroneutrality 电中性
73. fraction 分数;部分
74. transference 转移;转让
75. mole 摩尔
76. identical 相同的
77. notation 符号
78. galvanic cell 原电池
79. storage battery 蓄电池(组)
80. galvanic (流)电的;(电池)电流的,电镀的

81. convention 大会;惯例;约定;协定;习俗
82. cationic transport number 阳离子迁移系数
83. electrometallurgy 电冶金学
84. electrowinning 电解冶金法,电解沉积
85. speiss 黄渣
86. poling 插木还原
87. spontaneously 自然地;自发地;不由自主地
88. positively charged cation 带正电荷的阳离子
89. negatively charged anion 带负电荷的阴离子
90. batch 一次(配、制、装炉的)分量,一次操作所需的原料量

Exercises

1. What factors influenced the choice of extraction route?
2. What's the advantage of pyrometallurgy over hydrometallurgy and electrometallurgy?
3. Please give a brief overview of the advantages of hydrometallurgy in your own words.

READING & CRITICAL THINKING

LECTURE THREE
Nonmetals

Passage A An Overview of Nonmetals

In chemistry, a nonmetal is a chemical element that mostly lacks the characteristics of a metal. Physically, a nonmetal tends to have a relatively low melting point, boiling point, and density. A nonmetal is typically brittle when solid and usually has poor thermal conductivity and electrical conductivity. Chemically, nonmetals tend to have relatively high ionization energy, electron affinity, and electronegativity. They gain or share electrons when they react with other elements and chemical compounds. Seventeen elements are generally classified as nonmetals: most are gases (hydrogen, helium, nitrogen, oxygen, fluorine, neon, chlorine, argon, krypton, xenon and radon); one is a liquid (bromine); and a few are solids (carbon, phosphorus, sulfur, selenium, and iodine). Metalloids such as boron, silicon, and germanium are sometimes counted as nonmetals.

The nonmetals are divided into two categories reflecting their relative propensity to form chemical compounds: reactive nonmetals and noble gases. The reactive nonmetals vary in their nonmetallic character. The less electronegative of them, such as carbon and sulfur, mostly have weakness to moderately strong nonmetallic properties and tend to form covalent compounds with metals. The more electronegative of the reactive nonmetals, such as oxygen and fluorine, are characterised by stronger nonmetallic properties and a tendency to form predominantly ionic compounds with metals. The noble gases are distinguished by their great reluctance to form compounds with other elements.

The distinction between categories is not absolute. Boundary overlaps, including with the metalloids, occur as outlying elements in each category show or begin to show less-distinct, hybrid-like, or atypical properties.

Although five times more elements are metals than nonmetals, two of the nonmetals—hydrogen and helium—make up over 99 percent of the observable universe. Another nonmetal, oxygen, makes up almost half of the Earth's crust, oceans, and atmosphere. Living organisms are composed almost entirely of nonmetals: hydrogen, oxygen, carbon, and nitrogen. Nonmetals form more compounds than metals.

Definition and applicable elements

There is no rigorous definition of a nonmetal. Broadly, any element lacking a preponderance of metallic properties can be regarded as a nonmetal.

The elements generally classified as nonmetals include one element in Group 1 (hydrogen); one in Group 14 (carbon); two in Group 15 (nitrogen and phosphorus); three in Group 16 (oxygen, sulfur, and selenium); most of Group 17 (fluorine, chlorine, bromine, and iodine); and all of Group 18 (with the possible exception of oganesson).

As there is no widely agreed definition of a nonmetal, elements in the periodic table vicinity of where the metals meet the nonmetals are inconsistently classified by different authors. Elements sometimes also classified as nonmetals are the metalloids boron (B), silicon (Si), germanium (Ge), arsenic (As), antimony (Sb), tellurium (Te), and astatine (At). The nonmetal selenium (Se) is sometimes instead classified as a metalloid, particularly in environmental chemistry.

Properties

Nonmetals show more variability in their properties than metals do. These properties are largely determined by the interatomic bonding strengths and molecular structures of the nonmetals involved, both of which are subject to variation as the number of valence electrons in each nonmetal varies. Metals, in contrast, have more homogenous structures and their properties are more easily reconciled.

Physically, they largely exist as diatomic or monatomic gases, with the remainder having more substantial (open-packed) forms, unlike metals, which are nearly all solid and close-packed. If solid, they have a submetallic appearance (with the exception of sulfur) and are mostly brittle, as opposed to metals, which are lustrous, and generally ductile or malleable; they usually have lower densities than metals; are mostly poorer conductors of heat and electricity; and tend to have significantly lower melting points and boiling points than those of metals.

Chemically, the nonmetals mostly have high ionisation energies, high electron affinities (nitrogen and the noble gases have negative electron affinities) and high electronegativity values noting that, in general, the higher an element's ionisation energy, electron affinity, and electronegativity are, the more nonmetallic that element is. Nonmetals (including—to a limited extent—xenon and probably radon) usually exist as anions or oxyanions in aqueous solution; they generally form ionic or covalent compounds when combined with metals (unlike metals, which mostly form alloys with other metals); and have acidic oxides whereas the common oxides of nearly all metals are basic.

Complicating the chemistry of the nonmetals is the first row anomaly seen particularly in hydrogen, (boron), carbon, nitrogen, oxygen, and fluorine; and the alternation effect seen in (arsenic), selenium, and bromine. The first row anomaly largely arises from the electron configurations of the elements concerned.

Hydrogen is noted for the different ways it forms bonds. It most commonly forms covalent bonds. It can lose its single valence electron in aqueous solution, leaving behind a bare proton with tremendous polarising power. This subsequently attaches itself to the lone electron pair of an oxygen atom in a water molecule, thereby forming the basis of acid-base chemistry. Under certain conditions a hydrogen atom in a molecule can form a second, weaker bond with an atom or a group of atoms in another molecule. Such bonding, "helps give snowflakes their hexagonal symmetry, binds DNA into a double helix; shapes the three-dimensional forms of proteins; and even raises water's boiling point high enough to make a decent cup of tea".

Categories

Immediately, to the left of most nonmetals on the periodic table are metalloids such as boron, silicon, and germanium, which generally behave chemically like nonmetals, and are included here for comparative purposes. In this sense they can be regarded as the most metallic of nonmetallic elements.

Based on shared attributes, the nonmetals can be divided into two categories of reactive nonmetal and noble gas. The metalloids and the two nonmetal categories then span a progression in chemical nature from weakly nonmetallic, to moderately nonmetallic, to strongly nonmetallic (oxygen and the four nonmetallic halogens), and to almost inert. Analogous categories occur among the metals in the form of the weakly metallic (the post-transition metals), the moderately metallic (most of the transition metals), the strongly metallic (the alkali metal and alkaline earth metals, and the lanthanides and actinides), and the relatively inert (the noble transition metals).

As with categorisation schemes generally, there is some variation and overlapping of properties within and across each category. One or more of the metalloids are sometimes classified as nonmetals. Among the reactive nonmetals, carbon, phosphorus, selenium, and iodine—which border the metalloids—show some metallic character, as does hydrogen. Among the noble gases, radon is the most metallic and begins to show some cationic behaviour, which is unusual for a nonmetal.

Metalloid

The seven metalloids are boron (B), silicon (Si), germanium (Ge), arsenic (As), antimony (Sb), tellurium (Te), and astatine (At). On a standard periodic table, they occupy a diagonal area in the p-block extending from boron at the upper left to astatine at lower right, along the dividing line between metals and nonmetals shown on some periodic tables. They are called metalloids mainly in light of their physical resemblance to metals.

While they each have a metallic appearance, they are brittle and only fair conductors of electricity. Boron, silicon, germanium, and tellurium are semiconductors. Arsenic and antimony have the electronic band structures of semimetals although both have less stable semiconducting allotropes. Astatine has been predicted to have a metallic crystalline structure.

Chemically the metalloids generally behave like (weak) nonmetals. They have moderate ionisation energies, low to high electron affinities, moderate electronegativity values, are poor to moderately strong oxidising agents, and demonstrate a tendency to form alloys with metals.

Reactive nonmetal

The reactive nonmetals have a diverse range of individual physical and chemical properties. In periodic table terms they largely occupy a position between the weakly nonmetallic metalloids to the left and the noble gases to the right.

Physically, five are solids, one is a liquid (bromine), and five are gases. Of the solids, graphite carbon, selenium, and iodine are metallic-looking, whereas S_8 sulfur has a pale-yellow appearance. Ordinary white phosphorus has a yellowish-white appearance but the black allotrope, which is the most stable form of phosphorus, has a metallic-looking appearance. Bromine is a reddish-brown liquid. Of the gases, fluorine and chlorine are coloured pale yellow, and yellowish green. Electrically, most are insulators whereas graphite is a semimetal, and black phosphorus, selenium, and iodine are semiconductors.

Chemically, they tend to have moderate to high ionisation energies, electron affinities, and electronegativity values and be relatively strong oxidising agents. Collectively, the highest values of these properties are found among oxygen and the nonmetallic halogens. Manifestations of this status include oxygen's major association with the ubiquitous processes of corrosion and combustion and the intrinsically corrosive nature of the nonmetallic halogens. All five of these nonmetals exhibit a tendency to form predominately ionic compounds with metals whereas the remaining nonmetals tend to form predominately covalent compounds with metals.

Noble gas

Six nonmetals are categorised as noble gases: helium (He), neon (Ne), argon (Ar), krypton (Kr), xenon (Xe), and the radioactive radon (Rn). In periodic table terms they occupy the outermost right column. They are called noble gases in light of their characteristically very low chemical reactivity.

They have very similar properties, all being colorless, odorless, and nonflammable. With their closed valence shells, the noble gases have feeble interatomic forces of attraction resulting in very low melting and boiling points. That is why they are all gases under standard conditions, even those with atomic masses larger than many normally solid elements.

Chemically, the noble gases have relatively high ionization energies, negative electron affinities, and relatively high electronegativities. Compounds of the noble gases number is less than half a thousand, with most of these occurring via oxygen or fluorine combining with either krypton, xenon, or radon.

The status of the period 7 congener of the noble gases, oganesson (Og), is not known—it may or may not be a noble gas. It was originally predicted to be a noble gas but may instead be a fairly reactive solid with an anomalously low first ionisation potential, and a positive electron affinity, due to relativistic effects. On the other hand, if relativistic effects peak in period 7 at element 112, copernicium, oganesson may turn out to be a noble gas after all, albeit more reactive than either xenon or radon. While oganesson could be expected to be the most metallic of the group 18 elements, credible predictions on its status as either a metal or a nonmetal (or a metalloid) appear to be absent.

Terms

1. helium 氦
2. nitrogen 氮
3. fluorine 氟
4. neon 氖
5. chlorine 氯
6. argon 氩
7. krypton 氪
8. xenon 氙
9. radon 氡
10. bromine 溴
11. selenium 硒
12. boron 硼
13. germanium 锗
14. noble gas 稀有气体
15. reactive nonmetal 反应性非金属
16. metalloid 类金属
17. tellurium 碲
18. astatine 砹

Exercises

1. How do you define a nonmetal in chemistry?
2. How many categories can the nonmetals be divided into? What are they?
3. What nonmetals are categorized as noble gases?

Passage B Properties of Nonmetals (and Metalloids) by Group

Group 16

Oxygen is a colourless, odourless, and unpredictably reactive diatomic gas with a gaseous density of 1.429×10^{-3} g/cm^3 (marginally heavier than air). It is generally unreactive at room temperature. Thus, sodium metal will "retain its metallic lustre for days in the presence of absolutely dry air and can even be melted (m. p. 97.82 ℃) in the presence of dry oxygen without igniting". On the other hand, oxygen can react with many inorganic and organic compounds either spontaneously or under the right conditions, (such as a flame or a spark). It condenses to pale blue liquid at −82.962 ℃ and freezes into a light blue solid at −218.79 ℃. The solid form (density 0.0763 g/cm^3) has a cubic crystalline structure and is soft and easily crushed. Oxygen is an insulator in all of its forms. It has a high ionisation energy (1,313.9 kJ/mol), high electron affinity (141 kJ/mol), and high electronegativity (3.44). Oxygen is a strong oxidising agent ($O_2 + 4e^- \rightarrow 2H_2O, E^{\ominus} = 1.23$ V at pH 0). Metal oxides are largely ionic in nature.

Sulfur is a bright-yellow moderately reactive solid. It has a density of 2.07 g/cm^3 and is soft (MH 2.0) and brittle. It melts to a light yellow liquid at 95.3 ℃ and boils at 444.6 ℃. Sulfur has an abundance on the Earth one-tenth that of oxygen. It has an orthorhombic polyatomic (CN_2) crystalline structure and is brittle. Sulfur is an insulator with a band gap of 2.6 eV, and a photoconductor meaning its electrical conductivity increases a million-fold when illuminated. Sulfur has a moderate ionisation energy (999.6 kJ/mol), moderate electron affinity (200 kJ/mol), and high electronegativity (2.58). It is a poor oxidising agent ($S_8 + 2e^- \rightarrow H_2S, E^{\ominus} = 0.14$ V at pH 0). The chemistry of sulfur is largely covalent in nature, noting it can form ionic sulfides with highly electropositive metals. The common oxide of sulfur (SO_3) is strongly acidic.

Selenium is a metallic-looking, moderately reactive solid with a density of 4.81 g/cm^3 and is soft (MH 2.0) and brittle. It melts at 221 ℃ to a black liquid and boils at 685 ℃ to a dark yellow vapour. Selenium has a hexagonal polyatomic (CN_2) crystalline structure. It is a semiconductor with a band gap of 1.7 eV, and a photoconductor meaning its electrical conductivity increases a million-fold when illuminated. Selenium has a moderate ionisation energy (941.0 kJ/mol), high electron affinity (195 kJ/mol), and high electronegativity (2.55). It is a poor oxidising agent ($Se + 2e^- \rightarrow H_2Se, E^{\ominus} = -0.082$ V at pH 0). The chemistry of selenium is largely covalent in nature, noting it can form ionic selenides with highly electropositive metals. The common oxide of selenium (SeO_3) is strongly acidic.

Selenium is a chemical element with the symbol Se and atomic Number 34. It is a nonmetal (more rarely considered a metalloid) with properties that are intermediate between the elements above and below in the periodic table, sulfur, and tellurium and also has similarities to arsenic. It rarely occurs in its elemental state or as pure ore compounds in the Earth's crust. Selenium—from Ancient Greekσελήνη (selēnē) "Moon"—was discovered in 1817 by Jöns Jacob Berzelius, who noted the similarity of the new element to the previously discovered tellurium (named for the Earth).

Selenium is found in metal sulfide ores, where it partially replaces the sulfur. Commercially, selenium is produced as a byproduct in the refining of the seores, most often during production. Minerals that are pure selenide or selenate compounds are known but rare. The chief commercial uses for selenium today are glassmaking and pigments. Selenium is a semiconductor and is used in photocells. Applications in electronics, once important, have been mostly replaced with silicon semiconductor devices. Selenium is still used in a few types of DCpower surge protectors and one type of fluorescentquantum dot.

Selenium salts are toxic in large amounts, but trace amounts are necessary for cellular function in many organisms, including all animals. Selenium is an ingredient in many multivitamins and other dietary supplements, including infant formula. It is a component of the antioxidant enzymes glutathioneperoxidase and thioredoxin reductase (which indirectly reduce certain oxidised molecules in animals and some plants). It is also found in three deiodinase enzymes, which convert one thyroidhormone to another. Selenium requirements in plants differ by species, with some plants requiring relatively large amounts and others apparently requiring none.

Tellurium is a silvery-white, moderately reactive, and shiny solid, which has a density of 6.24 g/cm^3 and is soft (MH 2.25) and brittle. It is the softest of the commonly recognised metalloids. Tellurium reacts with boiling water, or when freshly precipitated even at 50 ℃, to give the dioxide and hydrogen: $Te + 2H_2O \rightarrow TeO_2 + 2H_2$. It has a melting point of 450 ℃ and a boiling point of 988 ℃. Tellurium has a polyatomic (CN_2) hexagonal crystalline structure. It is a semiconductor with a band gap of 0.32 to 0.38 eV. Tellurium has a moderate ionisation energy (869.3 kJ/mol), high electron affinity (190 kJ/mol), and moderate electronegativity (2.1). It is a poor oxidising agent ($Te + 2e^- \rightarrow H_2Te, E^\ominus = -0.45$ V at pH 0). The chemistry of tellurium is largely covalent in nature, noting it has an extensive organometallic chemistry and that many tellurides can be regarded as metallic alloys. The common oxide of tellurium (TeO_2) is amphoteric.

Tellurium is a chemical element with the symbol Te and atomic Number 52. It is a brittle, mildly toxic, rare, and silver-white metalloid. Tellurium is chemically related to selenium and sulfur, all three of which are chalcogens. It is occasionally found in the native form as elemental crystals. Tellurium is far more common in the universe as a whole than on Earth. Its extreme rarity in the Earth's crust, comparable to that of platinum, is due partly to its formation of a volatile hydride that caused tellurium to be lost to space as gas during the hot nebular formation of Earth, and partly to tellurium's low affinity for oxygen, which causes it to bind preferentially to other chalcophiles in dense minerals that sink into the core.

Tellurium-bearing compounds were first discovered in 1782 in a gold mine in

Kleinschlatten, Transylvania (now Zlatna, Romania) by Austrian mineralogist Franz-Joseph Müller von Reichenstein, although it was Martin Heinrich Klaproth who named the new element in 1798 after the Latin word for "earth", *tellus*. Gold telluride minerals are the most notable natural gold compounds. However, they are not a commercially significant source of tellurium itself, which is normally extracted as a by-product of copper and lead production.

Commercially, the primary use of tellurium is copper (tellurium copper) and steel alloys, where it improves machinability. Applications in CdTe solar panels and cadmium telluride semiconductors also consume a considerable portion of tellurium production. Tellurium is considered a technology-critical element.

Tellurium has no biological function, although fungi can use it in place of sulfur and selenium in aminoacids such as tellurocysteine and telluromethionine. In humans, tellurium is partly metabolized into dimethyl telluride, $(CH_3)_2Te$, a gas with a garlic-like odor exhaled in the breath of victims of tellurium exposure or poisoning.

Tellurium (Latin *tellus* meaning "earth") was discovered in the 18th century in a gold ore from the mines in Kleinschlatten, near today's city of Alba Iulia, Romania. This ore was known as "Faczebajer weiβes blättriges Golderz" (white leafy gold ore from Faczebaja, German name of Facebάnya, now Faţa Băii in Alba County) or antimonalischer Goldkies (antimonic gold pyrite), and according to Anton von Rupprecht, was Spieβglaskönig (argent molybdique), containing native antimony. In 1782, Franz-Joseph Müller von Reichenstein, who was then serving as the Austrian chief inspector of mines in Transylvania, concluded that the ore did not contain antimony but was bismuth sulfide. The following year, he reported that this was erroneous and that the ore contained mostly gold and an unknown metal very similar to antimony. After a thorough investigation that lasted three years and included more than fifty tests, Müller determined the specific gravity of the mineral and noted that when heated, the new metal gives off white smoke with a radish-like odor; that it imparts a red color to sulfuric acid; and that when this solution is diluted with water, it has a black precipitate. Nevertheless, he was not able to identify this metal and gave it the names aurum paradoxium (paradoxical gold) and metallum problem aticum (problem metal), because it did not exhibit the properties predicted for antimony.

In 1789, a Hungarian scientist, PάlKitaibel, discovered the element independently in an ore from Deutsch-Pilsen that had been regarded as argentiferous molybdenite, but later he gave the credit to Müller. In 1798, it was named by Martin Heinrich Klaproth, who had earlier isolated it from the mineral calaverite.

The 1960s brought an increase in thermoelectric applications for tellurium (as bismuth telluride), and in free-machining steel alloys, which became the dominant use.

Group 2

Fluorine is an extremely toxic and reactive pale yellow diatomic gas that, with a gaseous density of 1.696×10^{-3} g/cm^3, is about 40% heavier than air. Its extreme reactivity is such that it was not isolated (via electrolysis) until 1886 and was not isolated chemically until 1986. Its occurrence in an uncombined state in nature was first reported in 2012, but is contentious. Fluorine condenses to a pale yellow liquid at −188.11 ℃ and freezes into a colourless solid at −219.67 ℃. The solid form (density 1.7) has a cubic crystalline structure and is soft and easily crushed. Fluorine is an insulator in all of its forms. It has a high ionisation energy (1,681 kJ/mol), high electron affinity (328 kJ/mol), and high electronegativity (3.98). Fluorine is a powerful oxidising agent ($F_2 + 2e^- \rightarrow 2HF, E^{\ominus} = 2.87$ V at pH 0); "even water, in the form of steam, will catch fire in an atmosphere of fluorine". Metal fluorides are generally ionic in nature.

Chlorine is an irritating green-yellow diatomic gas that is extremely reactive, and has a gaseous density of 3.2×10^{-3} g/cm^3 (about 2.5 times heavier than air). It condenses at −34.04 ℃ to an amber-coloured liquid and freezes at −101.5 ℃ to a yellow crystalline solid. The solid form (density 1.9) has an orthorhombic crystalline structure and is soft and easily crushed. Chlorine is an insulator in all of its forms. It has a high ionisation energy (1,251.2 kJ/mol), high electron affinity (349 kJ/mol; higher than fluorine), and high electronegativity (3.16). Chlorine is a strong oxidising agent ($Cl_2 + 2e^- \rightarrow 2HCl, E^{\ominus} = 1.36$ V at pH 0). Metal chlorides are largely ionic in nature. The common oxide of chlorine (Cl_2O_7) is strongly acidic.

Bromine is deep brown diatomic liquid that is quite reactive, and has a liquid density of 3.1028 g/cm^3. It boils at 58.8 ℃ and solidifies at −7.3 ℃ to an orange crystalline solid (density 4.05). It is the only element, apart from mercury, known to be a liquid at room temperature. The solid form, like chlorine, has an orthorhombic crystalline structure and is soft and easily crushed. Bromine is an insulator in all of its forms. It has a high ionisation energy (1,139.9 kJ/mol), high electron affinity (324 kJ/mol), and high electronegativity (2.96). Bromine is a strong oxidising agent ($Br_2 + 2e^- \rightarrow 2HBr, E^{\ominus} = 1.07$ V at pH 0). Metal bromides are largely ionic in nature. The unstable common oxide of bromine (Br_2O_5) is strongly acidic.

Iodine, the rarest of the nonmetallic halogens, is a metallic looking solid that is moderately reactive, and has a density of 4.933 g/cm^3. It melts at 113.7 ℃ to a brown liquid and boils at 184.3 ℃ to a violet-coloured vapour. It has an orthorhombic crystalline structure with a flaky habit. Iodine is a semiconductor in the direction of its planes, with a band gap of about 1.3 eV and a conductivity of 1.7×10^{-8} S · cm^{-1} at room temperature. This is higher than selenium but lower than boron, the least electrically conducting of the recognised metalloids. Iodine is an insulator in the direction perpendicular to its planes. It has a high ionisation energy (1,008.4 kJ/mol), high electron affinity (295 kJ/mol), and high electronegativity (2.66). Iodine is a moderately strong oxidising agent ($I_2 + 2e^- \rightarrow 2I^-$, $E^{\ominus} = 0.53$ V at pH 0). Metal iodides are predominantly ionic in nature. The only stable oxide of iodine (I_2O_5) is strongly acidic.

Astatine is expected to have properties intermediate between iodine, a nonmetal with incident metallic properties, and tennessine, which is predicted to be a metal. Astatine has not so far been synthesised in sufficient quantities to enable a determination of its bulk properties. A macro-sized sample of astatine would vaporize itself due to radioactive heating; it is not known if such a phenomenon could be prevented with sufficient cooling. Many of the properties of astatine have nevertheless been predicted. It is expected to have a metallic appearance, a density of 6.35 ± 0.15 g/cm^3, a melting point of 302 ℃, a boiling point of 337 ℃, and a face-centred cubic crystalline structure. It has a moderate ionisation energy (899.003 kJ/mol), and is expected to have a high electron affinity (222 kJ/mol), and moderate electronegativity (2.2). Astatine is a weak oxidizing agent ($At + e^- \rightarrow At^-$, $E^{\ominus} = 0.3$ V at pH 0).

Terms

1. crystalline structure 晶体结构
2. insulator 绝缘体
3. ionisation energy 电离能
4. electron affinity 电子亲和力
5. electronegativity 电负性
6. density 密度
7. semiconductor 半导体
8. metalloid 类金属

Exercises

1. What are the properties of oxygen?
2. What is the chief commercial use of selenium?
3. What are the functions of tellurium?

READING & CRITICAL THINKING

LECTURE FOUR
Ferrous Metallurgy

Passage A Finery Forge

Finery forges, also called "refining furnace", is a kind of equipment for removing the impurity of the coarse metal in nonferrous metallurgy or iron-and-steel metallurgy. It can be used for ironmaking, where refining furnace is used to produce wrought iron from cast iron (in this context known as pig iron) by decarburization, and it also used for steelmaking. During the medieval period, means were found in Europe of producing wrought iron from cast iron using finery forges. In Europe, since the popularization of pig-iron smelting in the 14th century, the technology that the pig iron was used as the raw material for wrought iron and steel appeared. People put the charcoal in the uncovered stove, lit the fire, blew it up, and then added pig iron to the furnace and blew it together. When pig iron melts, it drops to the bottom of the furnace. In the process of dripping to the bottom of the furnace, oxygen in the air will burn off the carbon, silicon, manganese, phosphorus and other impurities it contains, so that the pig iron becomes wrought iron. If the carbon content can be properly controlled, it can also be fried into steel. The method of smelting pig iron for wrought iron or steel is called "wrought steel method" or "refining method ". This kind of furnace is called "wrought steel furnace" or "refining furnace". ① The finery forge process was replaced by the puddling process and the roller mill, both developed by Henry Cort in 1783 - 1784, but not becoming widespread until after 1800. ②

① 杨宽. 中国古代冶铁技术发展史[M]. 上海:上海人民出版社,2014.

② Ayres, Robert. Technological transformations and long waves[J]. *Technological Forecasting and Social Change*, 1990 (1): 1 -37.

History of finery forge

It is generally accepted that China is the first country in the world to invent the technology of smelting cast iron. According to the current archaeological evidence, the earliest ironware belonged to the late Spring and Autumn period. ① And the finery forge was used to refine wrought iron at least by the 3rd century BC in ancient China, based on the earliest archaeological specimens of cast and pig iron fined into wrought iron and steel found at the early Han Dynasty (202 BC - 220 AD) site at Tieshengguo. ② Pigott speculates that the finery forge existed in the previous Warring States period (403 - 221 BC), because of the wrought iron items from China dating to that period and there was no documented evidence of the bloomery ever being used in China. Wagner writes that in addition to the Han Dynasty hearths believed to be fining hearths, there is also pictoral evidence of the fining hearth from a Shandong tomb mural dated 1st to 2nd century AD, as well as a hint of written evidence in the 4th century AD Daoist text Taiping Jing. ③

In Europe, the concept of the finery forge may have been evident as early as the 13th century. However, it was perhaps not capable of being used to fashion plate armor until the 15th century, as described in conjunction with the waterwheel-powered blast furnace by the Florentine Italian engineer Antonio Averlino (c. 1400 - 1469). ④

The finery forge process began to be replaced in Europe from the late 18th century by others, of which puddling was the most successful, though some continued in use through the mid-19th century. The new methods used mineral fuel (coal or coke), and freed the iron industry from its dependence on wood to make charcoal.

Process

In the finery, a workman known as the "finer" remelted pig iron to oxidise the carbon (and silicon). This produced a lump of iron (with some slag) known as a bloom. This was consolidated using a water-powered hammer and returned to the finery.

---

① Ayres, Robert. Technological Transformations and Long Waves[J]. *Technological Forecasting and Social Change*, 1990, (1): 1 - 37.

② Pigott, Vincent C. *The Archaeometallurgy of the Asian Old World*[M]. Philadelphia: University of Pennsylvania Museum of Archaeology and Anthropology, 1999.

③ Wagner, Donald B. *The State and the Iron Industry in Han China*[M]. Copenhagen: Nordic Institute of Asian Studies Publishing, 2001: 80 - 83.

④ Williams, Alan R. *The Knight and the Blast Furnace: A History of the Metallurgy of Armor in the Middle Ages & the Early Modern Period* [M]. Leiden: Brill, 2003.

The next stages were undertaken by the "hammerman", who in some iron-making areas such as South Yorkshire was also known as the "stringsmith", who heated his iron in a string-furnace. Because the bloom is highly porous, and its open spaces are full of slag, the hammerman's or stringsmith's tasks were to beat (work) the heated bloom with a hammer to drive the molten slag out of it, and then to draw the product out into a bar to produce what was known as anconies or bar iron. To do this, he had to reheat the iron, for which he used the chafery. The fuel used in the finery had to be charcoal (later coke), as impurities in any mineral fuel would affect the quality of the iron.

Types of finery forges

Refining furnace is an indispensable equipment for purifying and removing impurities in metal smelting process. With the growth of social production and the development of smelting technology, refining furnaces are no longer only suitable for ironmaking, but are now more used for steelmaking, and a variety of refining methods have been derived at the same time. Extro-furnace refining is a common method in steelmaking, also called "ladle refining" or "secondary metallurgy". LF furnace and RH furnace are two main production equipment.

- **LF furnace**

LF furnace (ladle furnace) is a refining equipment developed in Japan in the early 1970s. Because of its simple equipment, low investment cost, flexible operation and good refining effect, it has become a rising star in the metallurgical industry and has been widely used and developed in Japan. LF refining is one of the main methods of secondary refining, and the key is to make white slag quickly. The purpose of LF slagging is to desulfurize, deoxidise, improve alloy yield and remove inclusions. The main tasks of LF furnace are: desulfurization; temperature regulation; accurate composition adjustment; improving the purity of molten steel and slagging.

* **Application of LF furnace**

LF furnace has the following uses:

(1) LF furnace is connected with electric furnace, which speeds up the production cycle of electric furnace and improves the quality of electric furnace steel.

(2) LF furnace connected with LD converter can reduce and refine converter steel, so the quality of steel can be improved.

(3) LF furnace can strictly adjust the composition and temperature of molten steel, which is beneficial to the hardenability of steel and the continuous casting of

special steel.

(4) LF furnace can heat and keep liquid steel warm and store liquid steel for a long time, which can ensure the smooth continuous casting. Therefore, it is an indispensable equipment for continuous casting workshop.

(5) LF furnace has the property of heat preservation of molten steel, which can be used to produce large ingots in a small furnace or to produce molten steel in one furnace into several spindles of different composition.

* **Development of LF furnace**

LF ladle refining furnace was first developed and used by Datong Iron and Steel Company of Japan. The reduction refining period of ARC furnace is replaced by LF smelting, which reduces the refining burden of ARC furnace and improves the productivity of it. LF furnace was only used to produce high-grade steel in the early stage of development. With the development of technology like metallurgy and continuous casting, the application range of LF furnace has been expanded. Since the LF furnace has the advantages of less investment, wide application and good refining effect, in recent years LF furnace has been used as the main secondary refining method at home and abroad.

- **RH furnace**

RH refining is called RH-vacuum degassing refining method. It was invented by German in 1959, in which RH was the first letter of two manufacturers adopting RH refining technology in Germany at that time. The application of vacuum technology in steelmaking began in 1952. At that time, slag often appeared in the pouring process when silicon steel with silicon content of about 2% was produced. After various tests, it was found that hydrogen and nitrogen in molten steel were the main reasons for slag formation, which could not be poured or rolled, and then various vacuum refining technologies began to appear, thus, the industrial scale vacuum treatment method of molten steel was created.

The advantages of molten steel treated by RH refining are obvious: the alloy does not react with slag; the alloy is directly added into the molten steel with high yield; the molten steel can be quickly and evenly mixed; the alloy composition can be controlled within a narrow range; the gas content is low, the inclusions are small, and the purity of molten steel is high. The top lance can also be used to adjust the temperature of chemical heating, so as to provide the molten steel with good fluidity, high purity and in line with the casting temperature for the continuous casting of multiple heats.

Otherwise, there are three types of refining furnace for copper-fire refining: reverberatory furnace, rotary furnace and tilting furnace.

(1) Reverberatory furnace

Reverberatory furnace is a traditional fire refining equipment, with simple structure, which is a kind of chamber furnace with surface heating. Easy to operate, it can process both cold materials and hot materials with solid fuel, liquid fuel, or gas fuel. With a changeable volume and furnace size ranging from small or large in fluctuations, reverberatory furnace is very adaptable in processing capacity from 1t to 400t. The factories processing more cold materials and the ones with small scales use the reverberatory furnace to produce anode copper more.

The reverberatory furnace also has the following disadvantages:

① Oxidation, reduction of air pipe, slagging-off and other operations are all manual operation. It needs a large amount of labor and high labor intensity. The working conditions are poor and it is difficult to achieve mechanization and automation.

② The air tightness of the furnace body is poor along with a large amount of heat dissipation loss, smoke leakage, poor workshop environment and other problems.

③ The amount of refractories and the flue pipe and auxiliary materials consume fast.

④ The stirring cycle of copper liquid in the furnace is poor and the operation efficiency is low.

(2) Rotary furnace

Rotary furnace is a fire refining equipment developed in the late 1950s. The furnace has a cylindrical furnace body for 360° rotation equipped with two to four air pipes, a furnace hole and a copper outlet. The tuyere is buried under the liquid by turning round the furnace body for oxidation and reduction operations. The rotary furnace body can be used for feeding in raw materials, slag discharge, copper discharge, simple and flexible in operation.

The rotary furnace has the following advantages:

① The furnace body structure is simple and has a high degree of mechanization and automation.

② The furnace volume varies from 100t to 500t with great processing capacity, good technical and economic indicators, and a high-labour productivity.

③ The operation of slag removal by sticking in air pipe has been canceled, and it reduces auxiliary material consumption.

④ The air tightness of the furnace body is better along with less heat dissipation loss, fuel consumption, and smoke leakage, and the workshop environment is comfortable.

(3) Tilting furnace

Tilting furnace was developed by the Swiss Melz furnace company in the mid-1960s. It was designed on the basis of reverberatory furnace and rotary furnace, drawing on the advantages of the above two furnace types. Its hearth is shaped like reverberatory furnace, maintaining a large heat exchange area and adopting the turnable way of rotary furnace. It adds fixed tuyere and removes air pipe and slag-removal operation, reducing labour intensity. Tilting furnace can process hot materials and cold ones, which is an ideal furnace type. The capacity of tilting furnace ranges from 55t to 350t. Although the tilting furnace is designed by integrating the functions of rotary furnace and reverberatory furnace, there are some shortcomings:

① The furnace shape is special, and the structure is complex, difficult in processing and high in investment.

② During operation, tilting the furnace body causes the center of gravity deviated, leaving the work in an unbalanced state, and the tilting mechanism has always been in a state of force.

③ The tilting of furnace body affects the stability of furnace roof and wall, especially after the furnace wall and furnace top have been burned.

④ When the furnace body tilts, the smoke outlet does not rotate with the furnace body. It is difficult in sealing and harmful to the environment.

Terms

1. finery forge 精炼炉
2. cast iron 生铁
3. decarburization 脱碳
4. fining hearth 精炉
5. puddling process 搅炼法
6. roller mill 滚压机,压延机
7. specimen 样品
8. bloomery 熟铁吹炼炉
9. blast furnace 鼓风炉
10. puddle 捣成泥浆;搅炼
11. mineral fuel 矿物燃料
12. hammer 铁锤
13. string-furnace 弦炉
14. impurity 杂质

15. extro-furnace refining 炉外精炼
16. secondary metallurgy 二次冶金
17. ladle furnace 钢包炉
18. metallurgical industry 冶金工业
19. desulfurize 使脱硫
20. electric furnace 电炉
21. converter 炼钢炉,吹风转炉
22. hardenability 可硬性,淬透性
23. spindle 纱锭
24. copper-fire refining 铜火精炼
25. reverberatory furnace 反射炉
26. rotary furnace 回转炉
27. tilting furnace 倾动炉
28. cold material 冷料
29. hot material 热料
30. solid fuel 固体燃料
31. liquid fuel 液态燃料
32. gas fuel 气体燃料
33. processing capacity 处理量
34. air pipe 风管
35. slagging-off 扒渣
36. air tightness 气密性
37. heat dissipation 散热
38. smoke leakage 烟雾泄漏
39. refractory 耐火材料
40. flue pipe 风管
41. auxiliary material 辅助材料
42. furnace hole 炉孔
43. copper outlet 铜出口
44. tuyere(冶金炉的)风口
45. raw material 原料
46. slag discharge 排渣
47. copper discharge 出铜
48. mechanization and automation 机械化和自动化
49. ARC furnace 电弧炉
50. RH-vacuum degassing refining method RH 真空脱气精炼法

Exercises

1. What is the definition of finery forge? Can you explain it in your own words?
2. What are the purposes of LF slagging?
3. When and how was the RH furnace invented?

Passage B Hot Blast

Definition

A hot blast refers to preheated air that is blown into a blast furnace. The technique is used in metallurgical operations, such as steelmaking, to increase the efficiency of the refining process. It works by recycling waste gases that would otherwise need to be vented. Burning these waste gases to preheat the air reduces the amount of fuel needed in the blast furnace and reduces the cost of operations.

The hot blast in a blast furnace, sometimes supplemented with natural gas, combust along with the coke to achieve the necessary temperature to process the ore. The preheating of this air improves combustion efficiency and reduces the amount of fuel required in the main blast furnace. It may also speed up the production of the final processed metal. This produces substantial cost savings by recycling energy that would otherwise be wasted.

As mentioned above, the hot blast stove is a thermal device that provides hot air for the process to complete the combustion process and heat transfer process. Its structure must include a combustion device for the fuel to burn in it and the airflow for heat transfer in it. For the regenerative hot blast stove that provides hot air to the blast furnace, it is necessary to have a combustion chamber and a burner to realize the combustion process, and a regenerator to stack the heat storage body that can complete the heat transfer process; to organize the airflow and realize the airflow process. The switching, the cold air chamber and various inlets and outlets and valves for air distribution are also indispensable. In addition, since the hot air required by the blast furnace has a certain pressure, a metal shell that can withstand the pressure is also essential. Therefore, the hot blast stove is a pressure vessel built with refractory materials in a metal shell.

Metallurgical processing typically makes use of a blast furnace to convert raw solids such as ore to purified liquid metals. High carbon fuel, known as coke, is used to heat the ore in order to reduce it to its constituent components so that purified metal may be separated out. Hot gases are produced as the by-product of this process.

Invention and spread

James Beaumont Neilson, previously foreman at Glasgow gas works, invented the system of preheating the blast for a furnace. He found that by increasing the temperature of the incoming air to 300 degrees Fahrenheit, he could reduce the fuel consumption from 8.06 tons to 5.16 tons with further reductions at even higher temperatures. He, with partners including Charles Macintosh, patented this in 1828. Initially, the heating vessel was made of wrought iron plates, but these were oxidised, and he substituted a cast iron vessel.

On the basis of a January 1828 patent, Thomas Botfield has a historical claim as the inventor of the hot blast method. Neilson is credited as the inventor of hot blast, because he won patent litigation. Neilson and his partners engaged in the substantial litigation to enforce the patent against in fringers. The spread of this technology across Britain was relatively slow. By 1840, 58 iron masters had taken out licenses, yielding a royalty income of £ 30,000 per year. By the time the patent expired there were 80 licenses. In 1843, just after it expired, 42 of the 80 furnaces in south Staffordshire were using hot blast, and uptake in south Wales was even slower. Other advantages of hot blast were that raw coal could be used instead of coke. In Scotland, the relatively poor "black band" ironstone could be profitably smelted. It also increased the daily output of furnaces. In the case of Calder ironworks from 5.6 tons per day in 1828 to 8.2 tons per day in 1833, which made Scotland the lowest cost steel-producing region in Britain in the 1830s.

Early hot blast stoves were troublesome, as thermal expansion and contraction could cause the breakage of pipes. This was somewhat remedied by supporting the pipes on rollers. It was also necessary to devise new methods of connecting the blast pipes to the tuyeres, as leather could not longer be used.

Ultimately, this principle was applied even more efficiently in regenerative heat exchangers, such as the Cowper stove (which preheat incoming blast air with waste heat from flue gas; these are used in modern blast furnaces), and in the open-hearth furnace (for making steel) by the Siemens-Martin process.

Independently, George Crane and David Thomas, of the Yniscedwyn Works in Wales, conceived of the same idea, and Crane filed for a British patent in 1836. They began producing iron by the new process on February 5, 1837. Crane subsequently bought Gessenhainer's patent and patented additions to it, controlling the use of the process in both Britain and the US. While Crane remained in Wales, Thomas moved to the US on behalf of the Lehigh Coal and Navigation Company and founded the Lehigh Crane Iron Company to utilize the process.

Types of hot blast

Through long-time practices, people have realized that only by using hot air as a medium and carrier can the heat utilization rate and heat work effect be greatly improved. Traditional electric heat sources and steam heat power are often equipped with multiple circulating fans during the transportation process, so that they will eventually indirectly form hot air for drying or heating operations. Obviously, this process has many shortcomings, such as a lot of waste of energy, excessive auxiliary equipment, and complicated process. The bigger problem is that this kind of heat source is helpless for those requiring higher temperature drying or baking operations. In response to these practical problems, after years of painstaking research, various hot blast stoves have been invented.

According to the operating principle, the hot blast can be divided into two types: regenerative reheating furnace and intermittent stove. Regenerative reheating furnace is a hot blast stove using checker bricks for the pebble stove. It consists of several parts: the shell, the combustion chamber, the checker work, and control valves and lines that regulate and deliver the various gases. The combustion chamber is arranged at the inside, top or outside. Therefore, there are three types of hot stoves: internal-combustion, top-combustion, and external-combustion. They have the same principle of operation. The size of the combustion chamber should be minimized so that the checker mass is as large as possible. The shell is welded steel cylinder 6 m to 9 m in diameter, and typically 20 m to 40 m high, and its insides are lined with refractory. According to the combustion method, it can be divided into top-combustion type, internal-combustion type, external-combustion type and so on. Increasing the hot air temperature of the hot blast is the key technology of BF ironmaking process. The industry has long studied how to increase the temperature of the hot air. Commonly used methods are to mix high-heating value gas, or increase the heat exchange of checker bricks in hot blast, or change the material and density of checker bricks, or change the shape of the regenerator (such as regenerative balls), and through various methods to preheat the gas and combustion air. Checker chamber with checker bricks in hot blast is currently the most commonly used form of hot blast in modern blast furnaces, especially large blast furnaces. The design of the stove burner is critical in assuring good combustion and efficient, stable operation. Internal ceramic burners with mixing capability of the gas and air streams are recommended to meet this requirement. The checker chamber is packed with checker bricks which provide many small, vertically-aligned flues for the high-temperatures gasses. The efficiency of the stove is improved as the surface area to volume ratio for the checker mass is increased. With high heat exchange temperature and high heat utilization rate. It is suitable for the production needs of large blast furnaces. But the disadvantages are also obvious. It is large in size and covers a large area with high purchase cost.

Except this type of stove, intermittent stoves mainly use heat-resistant heat exchangers as their core components. Metal heat exchangers cannot be used for this part, but only ceramic heat-resistant heat exchangers. The coal gas in the BF is fully burned in the combustion chamber. The hot air passes through the heat exchanger to exchange heat for fresh cold air, which can make the temperature of the fresh air reach more than 1,000 degrees. With high heat exchange temperature and high heat utilization rate, it is small in size and costs lower. However, the heat exchange temperature is not as high as the above mentioned one, and the scale of use is small. Modern blast furnaces mostly use regenerative reheating furnaces. Therefore, improving the heat transfer efficiency of hot blast stoves is of great significance to increasing the blast temperature. Increasing the heating area of checker bricks is one of the important technical measures to improve the heat transfer capacity.

Refractory materials

With the development of hot blast stoves, higher and higher requirements are put forward on the material and structure of refractory materials. Refractory materials and structures with different properties are selected according to the temperature and stress conditions of each part of the hot blast stove. It is particularly important that the refractory materials used in the high-temperature area are appropriate and the masonry structure is solid, which has a significant impact on the life and operational safety of the hot blast stove.

When selecting refractories for hot blast stoves, the quality indicators of refractory materials should be comprehensively evaluated, such as refractoriness, loading and melting point, creep resistance, compressive strength, spalling resistance, heat capacity, porosity. The high temperature area of the early hot blast stove used ordinary clay bricks. When the wind temperature is increased, the checker bricks of the checker chamber sink, the partition wall of the combustion chamber collapses and the wind blows, and the vault is cracked. Therefore, the quality of clay bricks was improved or alumina bricks of high quality were used, but the trouble that the settlement of checker bricks and the collapse of partition walls were not overcome.

Because clay bricks are cheap and easy to process, they are widely used in the middle and low-temperature areas of hot blast stoves. The refractory materials in the lower part of the combustion chamber and the upper part of the ceramic burner should be cordierite alumina bricks with good thermal stability. The hot blast stove also uses a large number of refractory insulation bricks. According to different insulation requirements, different temperature areas, and the corresponding refractory masonry materials, light silica bricks, light high alumina bricks and light clay bricks are selected. Various unshaped refractory materials are also used for the furnace shell. The refractory masonry of modern hot blast stove works under high temperature for a long time, and the temperature fluctuates periodically, the masonry produces uneven expansion, and thermal stress is generated inside the masonry.

Thermal stress is the main cause of the damage to the hot blast furnace body. Therefore, strict requirements are put forward for the design of hot blast stove masonry, and the following measures are taken:

(1) Adopt an independent structure for masonry with different working conditions;

(2) Reduce the size of the masonry in the high-temperature area, such as the radius of the vault of the external-combustion hot blast stove;

(3) Rider bricks are used for parts with poor conditions to improve the stability and integrity of the masonry structure;

(4) Handle expansion joints correctly to reduce the internal stress of the masonry;

(5) Strengthen heat insulation and try to make the temperature in the masonry uniform.

For key structures, such as the vault of hot blast stove, the finite element method is often used to calculate the stress changes accurately inside the vault to find a reasonable structure and determine the brick shape. Because at room temperature, the center line of the refractory masonry of the vault is aligned with the thrust line. When heated, the inside is expanded by the temperature difference between the inside and the outside, and the outside of the vault is stretched and separated from each other to concentrate the load on the inside. The force-bearing of the brick is reduced, and the local part is subject to extremely large compressive stress. When the crushing stress of the brick in the restrained state is exceeded, the brick will be crushed. At the same time, the vault itself will also be deformed, the top sinks, the brick joints in the middle crack outwards, and the thrust line in the vault moves to the lower end of the refractory bricks. Cracks are generated due to the alternating temperature in the masonry.

Operating principle

Due to the short life and low thermal efficiency, intermittent stoves are rarely used now. Regenerative reheating furnaces are more widely used. The following part mainly describes the working principle of the regenerative reheating furnace. The regenerative reheating furnace adopts a periodic working system, which is completed by combusting for storing heat and blasting for releasing heat. To keep the blast heated continuously, each blast furnace must be equipped with two or more hot blasts. The furnace is equipped with a set of valves for furnace replacement to make the hot blast furnace work alternately.

The equipment of hot blast stove is divided into two types: valves and devices for controlling combustion and doors for controlling blast. During the combusting period, the heating gas is sent to the combustion chamber through the pipe and the gas combustion valve, gas valve and gas flow regulation valve. The combustion air is combusted in the combustion chamber by the combustion-supporting fan through the air combustion valve and air flow regulation valve together with the gas. The flue gas flows upward, changes direction through the vault, and enters the checker chamber downward. In the checker chamber, the flue gas transfers heat to the checker bricks through radiation, convection and other heat transfer methods, so that the temperature of the bricks rises, and it gradually cools itself, and then the exhaust gas is discharged through the flue valve through the flue and chimney. After the checker bricks are heated and store enough heat, the combustion is stopped, the furnace is changed, and the air supply period is transferred. During the air supply period, the blast is sent into the hot blast stove through the cold blast pipe and the cold blast valve, conducts convective heat transfer with the heated brick from the bottom to the top and is heated. The heated blast is sent into the blast furnace by hot blast pipe and hot blast valve. However, at the beginning of the air supply period, the temperature of the blast sent out is high. At this time, a certain amount of cold air must be fed into the heated air through the air mixing valve and air mixing control valve to adjust the air temperature to meet the requirements. With the cooling of the bricks, the temperature of the hot air gradually decreases, and the air mixing control valve is gradually closed to reduce the amount of cold air mixed in. When the hot blast stove cannot maintain the specified blast temperature, the stove is changed again to convert the hot blast stove to the combusting period.

There are two methods of dealing with waste gases. One option is to clean them of any environmentally hazardous substances and vent them out of the system. However, this results in substantial waste of energy. Instead, waste gases may be routed to a special stove where they are burned to preheat air that is subsequently fed to the main blast furnace.

Rational allocation

Normal operation utilizes three stoves. One is always on blast, while the other two are on gas. When on gas, combustion air and clean blast furnace gas are introduced into the combustion chamber. The blast furnace gas may be enriched by either natural gas or coke oven gas as necessary. The turbulent mixing of the gas and air streams results in a short, intense flame after ignition. Flame temperatures 1,200℃ to 1,400℃. By the time the hot gasses have passed downward through the checker mass, the temperature of the gasses will have been reduced to 300℃ to 400℃ before being exhausted through the chimney valves. The stove is PLC controlled with the aim of dome temperature and waste gas temperature. In initial stage, the dome temperature is lower than the preset value, large flow rate of gas and air with reasonable fuel/air ratio is adopted for fast automation combustion. When the preset dome temperature is reached, then taking the dome temperature as target for auto control, i. e. keep the constant gas flow and enlarge the air flow. When the preset waste gas temperature is reached, then the operating mode auto change to let the waste gas temperature as target for auto control. When on blast, compressed cold blast air is forced upward through the checker chamber. A portion of the cold blast is bypassed around the stove and is reintroduced to the hot blast temperature. The mixer valve is open at the start of each cycle and closes progressively until the hot air leaving the stove is equal in temperature to the desired hot blast temperature. Further loss of temperature in the stove dictates changing to another stove and starting the next cycle. The refratory in stoves producing higher blast temperature obviously require higher overall stove operating temperatures. Higher alumina-content refractories can be used to raise stove dome temperatures to 1,315℃ safely, whereas further increases require the used of special creep-resistant silica refractories. Refractories in stoves must also resist thermal cycling and the effects of alkali and iron oxide. The life of refratory in stoves may be quite long, but is strongly dependent on the extent of contamination of the stove by impurities in the gas containing significant quantities of iron or alkali backdrafted through the stoves.

If a large blast furnace is equipped with 4 hot blast stoves, it can achieve staggered parallel air supply. Practice has proved that the air supply temperature can be increased by 20℃-50℃ without increasing the temperature of the vault of the hot blast. In the middle and late stages of the campaign, in the case of one hot blast stove overhaul, another three hot blast stoves can be used to work, so that the blast furnace production will be suspended. Many large blast furnaces at home and abroad have been equipped with 4 hot blast furnaces.

The use of 3 hot blast stoves can greatly reduce the construction investment and the floor space, which is also very practical. The adjustment and control of the air temperature is realized by mixed air, which also achieves the effect of high air temperature. This has become the trend of domestic and foreign small and medium blast furnace hot blast stove configuration.

Damage and repair of hot blast stove masonry

The most easily damaged parts of the hot blast stove masonry are the vault of the hot blast stove, the hot blast outlet, the upper part of the ceramic combustion chamber, and the combustion chamber partition and checker bricks of the internal-combustion hot blast stove. They are subject to high temperatures and rapid temperature changes. The use of refractory materials with good creep resistance improves the structure of each part of the masonry, so that the life of the hot blast stove can be extended. The distortion, disorderly sinking and collapse of checker bricks in the checker chamber have been overcome; the improvement of the combustion chamber partition structure and the use of external-combustion and top-combustion hot blast stoves have overcome the collapse of the partition wall and burning through short-circuit accidents that endanger the life of the hot blast stove.

It is necessary to further study the destructive effect of thermal stress during temperature changes and improve the structure of composite bricks. In operation, the vault and flue temperature must be strictly controlled to make the vault overheat and burn out the furnace pillars. Progress and improve the structure and material of the ceramic burner, so that the gas is evenly mixed, the flame length is shortened, the temperature is uniform, and the life of the upper part of the burner is prolonged. For the damage of the hot air outlet, hot air branch pipe and the upper part of the ceramic burner, the technology of thermal repair has also been developed. In the high-temperature state, the part to be repaired is isolated with a refractory heat shield, and then partially cooled, wearing protective clothing for partial repair and maintenance, so the life of the first generation of the hot blast stove has been extended to more than 30 years.

Extend the service life of the hot blast stove

The construction cost of the hot blast stove is very high, which is higher than the cost of the blast furnace body, about 1.1 - 1.5 times of it. The cost of overhaul is also very high, so it is very important to extend the life of the hot blast stove. The service life of the hot blast stove depends on the rationality of its design, including the form of the hot blast stove, the selection of refractory materials and its structure; the operation of the hot blast stove, including the oven, cooler and vault temperature, etc.; and maintenance and repair, etc.

Terms

1. hot blast furnace 热风炉
2. regenerative reheating furnace 蓄热式热风炉
3. intermittent stove 间断式热风炉
4. shell 炉壳
5. combustion chamber 燃烧室
6. internal-combustion furnace 内燃式炉
7. top-combustion furnace 顶燃式炉
8. external-combustion furnace 外燃式炉
9. gas combustion valve 煤气燃烧阀
10. gas valve 煤气阀
11. gas flow regulation valve 气体流量调节阀
12. combustion air 助燃空气
13. air combustion valve 空气燃烧阀
14. vault 拱顶
15. checker brick 格子砖
16. air supply period 送风期
17. cold blast pipe 冷风阀
18. cold blast valve 冷风管道
19. hot blast pipe 热风阀
20. hot blast valve 热风管道
21. air mixing valve 混风阀
22. air mixing control valve 混风调节阀
23. masonry structure 砌体结构
24. settlement 沉降
25. thermal stress 热应力
26. campaign 炉龄期,炉子使用期
27. air flow regulation valve 空气流量调节阀

Exercises

1. How many types of hot blast stoves are there nowadays?
2. How many parts does the hot blast stove consist of?
3. Can you describe the operating principles of Cowper stove?
4. Where can the combustion chamber be arranged?

Passage C An Overview of Blast Furnaces

A blast furnace is a type of metallurgical furnace used for smelting to produce industrial metals, generally pig iron, and others such as lead or copper. Blast refers to the combustion air being "forced" or supplied above atmospheric pressure. In a blast furnace, fuel (coke), ores, and flux (limestone) are continuously supplied through the top of the furnace.

While a hot blast of air (sometimes with oxygen enrichment) is blown into the lower section of the furnace through a series of pipes called tuyeres, so that the chemical reactions take place throughout the furnace as the material falls downward. The end products are usually molten metal and slag phases tapped from the bottom, and waste gases (flue gas) exiting from the top of the furnace. The downward flow of the ore along with the flux in contact with an upflow of hot, carbon monoxide-rich combustion gases is a countercurrent exchange and chemical reaction process. During blast furnace production, iron ore, coke, and flux (limestone) for slagging are loaded from the top of the furnace, and preheated air is blown in from the tuyere located at the lower part of the furnace along the periphery of the furnace. The molten iron produced is discharged from the iron hole. The unreduced impurities in the iron ore combine with fluxes such as limestone to form slag, which is discharged from the slag port. The generated gas is discharged from the top of the furnace, and after dust removal, it is used as fuel for hot blast stoves, heating furnaces, coke ovens, boilers, etc. The main product of blast furnace smelting is pig iron, as well as by-products of blast furnace slag and blast furnace gas.

In contrast, air furnaces (such as reverberatory furnaces) are naturally aspirated, usually by the convection of hot gases in a chimney flue. According to this broad definition, bloomeries for iron, blowing houses for tin, and smelt mills for lead would be classified as blast furnaces. However, the term has usually been limited to those used for smelting iron ore to produce pig iron, an intermediate material used in the production of commercial iron and steel, and the shaft furnaces used in combination with sinter plants in base metals smelting.

Nowadays, the blast furnace uses steel plates as the furnace shell, which is lined with refractory bricks. From top to bottom, the blast furnace body is divided into 5 parts: throat, body, waist, belly and hearth. Due to the good technical and economic indicators of blast furnace ironmaking, simple process, large production, high labor productivity and low energy consumption, etc., the iron was produced by this method accounts for most of the world's total iron output.

Terms

1. metallurgical 冶金的
2. pig iron 生铁
3. ore 矿石
4. end product 制成品
5. slag phase 渣相
6. monoxide 一氧化物
7. iron ore 铁矿石
8. slag port 渣口
9. by-product 副产品
10. refractory brick 耐火砖
11. shaft furnace 竖炉
12. sinter plant 烧结厂
13. base metal 贱金属
14. belly 腹部
15. hearth 炉

Exercises

1. How does a blast furnace operate?

2. As the material falls downward, what kind of chemical reactions take place throughout the furnace?

Passage D The History of Blast Furnaces

Cast iron has been found in China dating back to the 5th century BC, but the earliest extant blast furnaces in China date back to the 1st century AD and in the West from the High Middle Ages. They spread from the region around Namur in Wallonia (Belgium) in the late 15th century, being introduced to England in 1491. The fuel used in these was invariably charcoal. The successful substitution of coke for charcoal is widely attributed to English inventor Abraham Darby in 1709. The efficiency of the process was further enhanced by the practice of preheating the combustion air (hot blast), patented by Scottish inventor James Beaumont Neilson in 1828.

Archaeological evidence shows that bloomeries appeared in China around 800 BC. Originally it was thought that the Chinese started casting iron right from the beginning, but this theory has been debunked by the discovery of more than ten iron digging implements found in the tomb of Duke Jing of Qin (d. 537 BC), whose tomb is located in Fengxiang County, Shanxi (a museum exists on the site today). There is however no evidence of the bloomery in China after the appearance of the blast furnace and cast iron. In China, blast furnaces produced cast iron, which was then either converted into finished implements in a cupola furnace, or turned into wrought iron in a fining hearth.

Although cast iron farm tools and weapons were widespread in China by the 5th century BC, employing workforces of over 200 men in iron smelters from the 3rd century onward, the earliest blast furnaces constructed were attributed to the Han Dynasty in the 1st century AD. These early furnaces had clay walls and used phosphorus-containing minerals as a flux. Chinese blast furnaces ranged from around two to ten meters in height, depending on the region. The largest ones were found in modern Sichuan and Guangdong, while the "dwarf" blast furnaces were found in Dabieshan. In construction, they are both around the same level of technological sophistication.

The effectiveness of the Chinese blast furnace was enhanced during this period by the engineer Du Shi, who applied the power of waterwheels to piston-bellows in forging cast iron. Donald Wagner suggests that early blast furnace and cast-iron production evolved from furnaces used to melt bronze. Certainly, though, iron was essential to military success by the time the State of Qin had unified China (221 BC). Usage of the blast and cupola furnace remained widespread during the Song and Tang Dynasties. By the 11th century, the Song Dynasty Chinese iron industry made a switch of resources from charcoal to coke in casting iron and steel, sparing thousands of acres of woodland from felling. This may have happened as early as the 4th century AD.

The primary advantage of the early blast furnace was in large-scale production and making iron implements more readily available to peasants. Cast iron is more brittle than wrought iron or steel, which required additional fining and then cementation or co-fusion to produce, but for menial activities such as farming it sufficed. By using the blast furnace, it was possible to produce larger quantities of tools such as ploughshares more efficiently than the bloomery. In areas where quality was important, such as warfare, wrought iron and steel were preferred. Nearly all Han period weapons are made of wrought iron or steel, with the exception of axe-heads, of which many are made of cast iron.

Blast furnaces were also later used to produce gunpowder weapons such as cast-iron bomb shells and cast-iron cannons during the Song dynasty. Also, during the Yuan dynasty, the blower also began to be innovated. During the Ming dynasty, with the emergence of new wooden bellows, we witnessed the rapid development of small blast furnaces. This is the process and result of the early development of China's blast furnace.

Medieval Europe

The simplest forge, known as the Corsican, was used prior to the advent of Christianity. Examples of improved bloomeries are the Stückofen (sometimes called wolf-furnace), which remained until the beginning of the 19th century. Instead of using natural draught, air was pumped in by a trompe, resulting in better quality iron and an increased capacity. This pumping of airstream in with bellows is known as cold blast, and it increases the fuel efficiency of the bloomery and improves yield. They can also be built bigger than natural draught bloomeries.

Oldest European blast furnaces

The oldest known blast furnaces in the West were built in Dürstel in Switzerland, the Märkische Sauerland in Germany, and at Lapphyttan in Sweden, where the complex was active between 1205 and 1300. At Noraskog in the Swedish parish of Järnboås, there has also been found traces of blast furnaces dating even earlier, possibly to around 1100. These early blast furnaces, like the Chinese examples, were very inefficient compared to those used today. The iron from the Lapphyttan complex was used to produce balls of wrought iron known as osmonds, and these were traded internationally—a possible reference occurs in a treaty with Novgorod from 1203 and several certain references in accounts of English customs from the 1250s and 1320s. Other furnaces of the 13th to 15th centuries have been identified in Westphalia.

The technology required for blast furnaces may have either been transferred from China, or may have been an indigenous innovation. Al-Qazvini in the 13th century and other travellers subsequently noted an iron industry in the Alburz Mountains to the south of the Caspian Sea. This is close to the silk route, so that the use of technology derived from China is conceivable. Much later descriptions record blast furnaces about three metres high. As the Varangian Rus' people from Scandinavia traded with the Caspian (using their Volga trade route), it is possible that the technology reached Sweden by this means. The step from bloomery to true blast furnace is not big. Simply just building a bigger furnace and using bigger bellows to increase the volume of the blast and hence the amount of oxygen leads inevitably into higher temperatures, bloom melting into liquid iron, and cast iron flowing from the smelters. Already the Vikings are known to have used double bellows, which greatly increases the volumetric flow of the blast.

The Caspian region may also have been the source for the design of the furnace at Ferriere, described by Filarete, involving a water-powered bellows at Semogo [it] in Valdidentro in northern Italy in 1226 in a two-stage process. With this process, the molten iron was tapped twice a day into water, thereby granulating it.

Terms

1. charcoal 木炭	2. preheat 预热
3. archaeological 考古的	4. cupola furnace 化铁炉,冲天炉
5. clay 黏土	6. mineral 矿物
7. piston-bellow 活塞波纹管	8. acre 英亩
9. cementation 黏固	10. ploughshare 铧
11. axe-head 轴头	12. cannon 大炮
13. shell 炮弹	14. pump in 用泵吸入
15. trompe 角拱	16. bellows 风箱
17. volumetric 体积的	18. granulate 使成颗粒

Exercises

1. How did China's blast furnace develop during ancient times?
2. What are the primary advantage of the early blast furnace?

Passage E The Origin and Spread of Early Modern Blast Furnaces

Due to the casting of cannon, the blast furnace came into widespread use in France in the mid-15th century.

The direct ancestor of these used in France and England was in the Namur region in what is now Wallonia (Belgium). From there, they spread first to the Pays de Bray on the eastern boundary of Normandy and from there to the Weald of Sussex, where the first furnace (called Queenstock) in Buxted was built in about 1491, followed by one at Newbridge in Ashdown Forest in 1496. They remained few in number until about 1530 but many were built in the following decades in the Weald, where the iron industry perhaps reached its peak about 1590. Most of the pig iron from these furnaces was taken to finery forges for the production of bar iron.

The first British furnaces outside the Weald appeared during the 1550s, and many were built in the remainder of that century and the following ones. The output of the industry probably peaked about 1620, and was followed by a slow decline until the early 18th century. This was apparently because it was more economic to import iron from Sweden and elsewhere than to make it in some more remote British locations. Charcoal that was economically available to the industry was probably being consumed as fast as the wood to make it grew. The Backbarrow blast furnace built in Cumbria in 1711 has been described as the first efficient example.

The first blast furnace in Russia opened in 1637 near Tula and was called the Gorodishche Works. The blast furnace spread from here to central Russia and then finally to the Urals.

Coke blast furnaces

In 1709, at Coalbrookdale in Shropshire, England, Abraham Darby began to fuel a blast furnace with coke instead of charcoal. Coke's initial advantage was its lower cost, mainly because making coke required much less labor than cutting trees and making charcoal, but using coke also overcame localized shortages of wood, especially in Britain and on the Continent. Metallurgical grade coke will bear heavier weight than charcoal, allowing larger furnaces. A disadvantage is that coke contains more impurities than charcoal, with sulfur being especially detrimental to the iron's quality. Coke's impurities were more of a problem before hot blast reduced the amount of coke required and before furnace temperatures were hot enough to make slag from limestone free flowing. (Limestone ties up sulfur. Manganese may also be added to tie up sulfur).

Coke iron was initially only used for foundry work, making pots and other cast iron goods. Foundry work was a minor branch of the industry, but Darby's son built a new furnace at nearby Horsehay, and began to supply the owners of finery forges with coke pig iron for the production of bar iron. Coke pig iron was by this time cheaper to produce than charcoal pig iron. The use of a coal-derived fuel in the iron industry was a key factor in the British Industrial Revolution. Darby's original blast furnace has been archaeologically excavated and can be seen in situ at Coalbrookdale, part of the Ironbridge Gorge Museums. Cast iron from the furnace was used to make girders for the world's first iron bridge in 1779. The Iron Bridge crosses the River Severn at Coalbrookdale and remains in use for pedestrians.

Steam-powered blast

The steam engine was applied to power blast air, overcoming a shortage of water power in areas where coal and iron ore were located. The cast iron blowing cylinder was developed in 1768 to replace the leather bellows, which wore out quickly. The steam engine and cast-iron blowing cylinder led to a large increase in British iron production in the late 18th century.

Hot blast

Hot blast was the single most important advance in fuel efficiency of the blast furnace and was one of the most important technologies developed during the Industrial Revolution. Hot blast was patented by James Beaumont Neilson at Wilsontown Ironworks in Scotland in 1828. Within a few years of the introduction, hot blast was developed to the point where fuel consumption was cut by one-third using coke or two-thirds using coal, while furnace capacity was also significantly increased. Within a few decades, the practice was to have a "stove" as large as the furnace next to it into which the waste gas (containing CO) from the furnace was directed and burnt. The resultant heat was used to preheat the air blown into the furnace.

Hot blast enabled the use of raw anthracite coal, which was difficult to light, to the blast furnace. Anthracite was first tried successfully by George Crane at Ynyscedwyn Ironworks in south Wales in 1837. It was taken up in America by the Lehigh Crane Iron Company at Catasauqua, Pennsylvania, in 1839. Anthracite use declined when very high capacity blast furnaces requiring coke were built in the 1870s.

Modern furnaces

- **Iron blast furnaces**

The blast furnace remains an important part of modern iron production. Modern furnaces are highly efficient, including cowper stoves to preheat the blast air and employ recovery systems to extract the heat from the hot gases exiting the furnace. Competition in industry drives higher production rates. The largest blast furnace in the world is in Republic of Korea, with a volume around 6,000 m^3. It can produce around 5,650,000 tonnes of iron per year.

This is a great increase from the typical 18th-century furnaces, which averaged about 360 tonnes (350 long tons; 400 short tons) per year. Variations of the blast furnace, such as the Swedish electric blast furnace, have been developed in countries which have no native coal resources.

- **Lead blast furnaces**

Blast furnaces are currently rarely used in copper smelting, but modern lead smelting blast furnaces are much shorter than iron blast furnaces and are rectangular in shape. The overall shaft height is around 5m to 6m. Modern lead blast furnaces are constructed using water-cooled steel or copper jackets for the walls, and have no refractory linings in the side walls. The base of the furnace is a hearth of refractory material (bricks or castable refractory). Lead blast furnaces are often open-topped rather than having the charging bell used in iron blast furnaces.

The blast furnace used at the Nyrstar Port Pirie lead smelter differs from most other lead blast furnaces in that it has a double row of tuyeres rather than the single row normally used. The lower shaft of the furnace has a chair shape with the lower part of the shaft being narrower than the upper. The lower row of tuyeres being located in the narrow part of the shaft. This allows the upper part of the shaft to be wider than the standard.

- **Zinc blast furnaces (imperial smelting furnaces)**

The blast furnaces used in the imperial smelting process (ISP) were developed from the standard lead blast furnace, but are fully sealed. This is because the zinc produced by these furnaces is recovered as metal from the vapor phase, and the presence of oxygen in the off-gas would result in the formation of zinc oxide. Blast furnaces used in the ISP have a more intense operation than standard lead blast furnaces, with higher air blast rates of hearth area and a higher coke consumption. Zinc production with the ISP is more expensive than with electrolytic zinc plants, so several smelters operating this technology have closed in recent years. However, ISP furnaces have the advantage of being able to treat zinc concentrates containing higher levels of lead than electrolytic zinc plants.

Here, we list several different blast furnaces to illustrate the origin and the spread of early modern blast furnaces. We can see the development of early modern blast furnaces. In the following passages, we will discuss modern process of how modern blast furnaces operate.

Terms

1. bar iron 铁条
2. Gorodishche works 戈罗迪什作品
3. archaeologically 考古学地
4. girder 大梁
5. cylinder 气缸
6. anthracite 无烟煤
7. rectangular 矩形的，长方形的
8. shaft 轴
9. lining 衬里；内衬
10. zinc 锌
11. vapor phase 气相

Exercises

1. How did the British Industrial Revolution relate to the iron industry?
2. Can you briefly describe the characteristics of early modern blast furnaces?

Passage F Modern Furnaces

Modern furnaces are equipped with an array of supporting facilities to increase efficiency, such as ore storage yards where barges are unloaded. The raw materials are transferred to the stockhouse complex by ore bridges, or rail hoppers and ore transfer cars. Rail-mounted scale cars or computer-controlled weight hoppers weigh out the various raw materials to yield the desired hot metal and slag chemistry. The raw materials are brought to the top of the blast furnace via a skip car powered by winches or conveyor belts.

There are different ways in which raw materials are charged into the blast furnace. Some blast furnaces use a "double bell" system where two "bells" are used to control the entry of raw materials into the blast furnace. The purpose of the two bells is to minimize the loss of hot gases in the blast furnace. First, raw materials are emptied into the upper or small bell which then opens to empty the charge into the large bell. The small bell then closes, to seal the blast furnace, while the large bell rotates to provide specific distribution of materials before dispensing the charge into the blast furnace. A more recent design is to use a "bell-less" system. These systems use multiple hoppers to contain each raw material, which is then discharged into the blast furnace through valves. These valves are more accurate at controlling how much of each constituent is added, as compared to the skip or conveyor system, thereby increasing the efficiency of the furnace. Some of these bell-less systems also implement a discharge chute in the throat of the furnace (as with the Paul Wurth top) to precisely control where the charge is placed.

The iron making blast furnace itself is built in the form of a tall structure, lined with refractory brick, and profiled to allow for expansion of the charged materials as they heat during their descent, and subsequent reduction in size as melting starts to occur. Coke, limestone flux, and iron ore (iron oxide) are charged into the top of the furnace in a precise filling order which helps control gas flow and the chemical reactions inside the furnace. Four "uptakes" allow the hot, dirty gas high in carbon monoxide content to exit the furnace throat, while "bleeder valves" protect the top of the furnace from sudden gas pressure surges. The coarse particles in the exhaust gas settle in the "dust catcher" and are dumped into a railroad car or truck for disposal, while the gas itself flows through a Venturi scrubber and/or electrostatic precipitators and a gas cooler to reduce the temperature of the cleaned gas.

The "cast-house" at the bottom half of the furnace contains the bustle pipe, water cooled copper tuyeres and the equipment for casting the liquid iron and slag. Once a "taphole" is drilled through the refractory clay plug, liquid iron and slag flow down a trough through a "skimmer" opening, separating the iron and slag. Modern, larger blast furnaces may have as many as four tapholes and two casthouses. Once pig iron and slag has been tapped, the taphole is again plugged with refractory clay.

The tuyeres are used to implement a hot blast, which is used to increase the efficiency of the blast furnace. The hot blast is directed into the furnace through water-cooled copper nozzles called tuyeres near the base. The hot blast temperature can be from 900 ℃ to 1,300 ℃ (1,652 ℉ to 2,372 ℉) depending on the stove design and condition. The temperatures they deal with may be 2,000 ℃ to 2,300 ℃ (3,632 ℉ to 4,172 ℉). Oil, tar, natural gas, powdered coal and oxygen can also be injected into the furnace at a tuyere level to combine with the coke to release additional energy and increase the percentage of reducing gases present which is necessary to increase productivity.

Take China as an example, "three transmissions and one anti" is the disciplinary basis of process industry and process engineering, which provides an important theoretical guarantee for the modernization of process engineering research and design. In recent years, due to the development characteristics of China's process industry, resource and environmental problems have become prominent. To better solve energy and environmental problems, the complementary and synergistic thinking of "three transmissions and one anti" and "three loops and one network" is proposed, hoping to fundamentally solve the resource, energy and environmental problems in the process industry. The so-called three rings mean that new processes and technologies meet the requirements of the environment (first ring); material and energy realize the circulation within the industry and equipment (second ring); resources and energy realize the whole society, cross-industry, and cross-regional circulation (third ring) utilization and matching. One network is to build a coal-based clean gas energy network to achieve self-sufficiency in clean energy and ensure China's energy security. China's energy is mainly coal, and the direct use of coal has caused energy waste, air pollution, and water pollution. However, as an energy source that has to be used, only clean and efficient conversion is fundamental. Blast furnaces are important equipment in the iron and steel industry. In recent years, due to the overcapacity of crude steel, the recycling of scrap steel, and the difficulty of repaying capital investment in infrastructure, some blast furnaces are facing the risk of demolition, which will inevitably lead to the waste of fixed assets. Therefore, how to find a way out for large and medium blast furnaces is a problem. In addition to ironmaking, the energy conversion and solid waste conversion functions of blast furnaces have not yet come into play. For this reason, it is proposed to transform the blast furnace into a gas generator, use China's abundant noncoking coal briquette to gasify in the blast furnace, reuse gas purification technology, waste heat recovery technology, realize the clean conversion of coal, and finally form a clean gas energy network. China hopes to help the process industry, especially the energy industry and heavy industry, with the complementary concepts of "three transmissions and one anti" and "three loops and one network". Make China's iron and steel industry develop from the first in production to the improvement of equipment functions, equipment upgrades, product structure expansion, etc., to achieve industrial upgrading, break through industry barriers, realize the transformation of excess steel to coal-based clean energy, and help the process industry's "curve" overtake.

Terms

1. barge 驳船
2. stockhouse 仓库
3. hopper 料斗
4. winch 绞车
5. valve 阀门
6. chute 溜槽
7. descent 下降
8. catcher 捕捉者;接手
9. Venturi scrubber 文丘里洗涤器
10. bustle pipe 环形风管
11. taphole 塞孔
12. clay plug 黏土塞
13. casthouse 出铁场
14. copper nozzle 铜喷嘴
15. tar 焦油

Exercises

1. How does modern furnaces try to increase their efficiency?
2. How does China try to regenerate modern blast furnace?

Passage G Process Engineering and Chemistry

Blast furnaces operate on the principle of chemical reduction whereby carbon monoxide, having a stronger affinity for the oxygen in iron ore than iron does, reduces the iron to its elemental form. Blast furnaces differ from bloomeries and reverberatory furnaces in that in a blast furnace, flue gas is in direct contact with the ore and iron, allowing carbon monoxide to diffuse into the ore and reduce the iron oxide to elemental iron mixed with carbon. The blast furnace operates as a countercurrent exchange process whereas a bloomery does not. Another difference is that bloomeries operate as a batch process while blast furnaces operate continuously for long periods because they are difficult to start up and shut down. (See: continuous production) Also, the carbon in pig iron lowers the melting point below that of steel or pure iron; in contrast, iron does not melt in a bloomery.

Silica has to be removed from pig iron. It reacts with calcium oxide (burned limestone) and forms a silicate which floats to the surface of the molten pig iron as "slag". Historically, to prevent contamination from sulfur, the best-quality iron was produced with charcoal.

The downward moving column of ore, flux, coke or charcoal and reaction products must be porous enough for the flue gas to pass through. This requires the coke or charcoal to be in large enough particles to be permeable, which means there cannot be an excess of fine particles. Therefore, the coke must be strong enough so it will not be crushed by the weight of the material above it. Besides physical strength of the coke, it must also be low in sulfur, phosphorus, and ash. This necessitates the use of metallurgical coal, which is a premium grade due to its relative scarcity.

The main chemical reaction producing the molten iron is:

$$Fe_2O_3 + 3CO \rightarrow 2Fe + 3CO_2$$

This reaction might be divided into multiple steps, with the first being that preheated blast air blown into the furnace reacts with the carbon in the form of coke to produce carbon monoxide and heat:

$$2C(s) + O_2(g) \rightarrow 2CO(g)$$

The hot carbon monoxide is the reducing agent for the iron ore and reacts with the iron oxide to produce molten iron and carbon dioxide. Depending on the temperature in the different parts of the furnace (warmest at the bottom), the iron is reduced in several steps. At the top, where the temperature usually is in the range between 200 ℃ and 700 ℃, the iron oxide is partially reduced to iron(II,III) oxide, Fe_3O_4:

$$3Fe_2O_3(s) + CO(g) \rightarrow 2Fe_3O_4(s) + CO_2(g)$$

At temperatures around 850 ℃, further down in the furnace, the iron (II,III) is reduced further to iron(II) oxide:

$$Fe_3O_4(s) + CO(g) \rightarrow 3FeO(s) + CO_2(g)$$

Hot carbon dioxide, unreacted carbon monoxide, and nitrogen from the air pass up through the furnace as fresh feed material travels down into the reaction zone. As the material travels downward, the counter-current gases both preheat the feed charge and decompose the limestone to calcium oxide and carbon dioxide:

$$CaCO_3(s) \rightarrow CaO(s) + CO_2(g)$$

The calcium oxide formed by decomposition reacts with various acidic impurities in the iron (notably silica), to form a fayalitic slag which is essentially calcium silicate, CaSiO:

$$SiO_2 + CaO \rightarrow CaSiO_3$$

As the iron(II) oxide moves down to the area with higher temperatures, ranging up to 1,200 ℃ degrees, it is reduced further to iron metal:

$$FeO(s) + CO(g) \rightarrow Fe(s) + CO_2(g)$$

The carbon dioxide formed in this process is re-reduced to carbon monoxide by the coke:

$$C(s) + CO_2(g) \rightarrow 2CO(g)$$

The temperature-dependent equilibrium controlling the gas atmosphere in the furnace is called the Boudouard reaction:

$$2CO \rightleftharpoons CO_2 + C$$

The "pig iron" produced by the blast furnace has a relatively high carbon content of around 4%—5% and usually contains too much sulphur, making it very brittle, and of limited immediate commercial use. Some pig iron is used to make cast iron. The majority of pig iron produced by blast furnaces undergoes further processing to reduce the carbon and sulphur content and produces various grades of steel used for construction materials, automobiles, ships, and machinery. Desulphurisation usually takes place during the transport of the liquid steel to the steelworks. This is done by adding calcium oxide, which reacts with the iron sulfide contained in pig iron to form calcium sulfide (called lime desulfurization). In a further process step, the so-called basic oxygen steelmaking, the carbon is oxidised by blowing oxygen onto the liquid pig iron to form crude steel.

Although the efficiency of blast furnaces is constantly evolving, the chemical process inside the blast furnace remains the same. According to the American Iron and Steel Institute: "Blast furnaces will survive into the next millennium because the larger, efficient furnaces can produce hot metal at costs competitive with other iron-making technologies." One of the biggest drawbacks of the blast furnaces is the inevitable carbon dioxide production as iron is reduced from iron oxides by carbon and as of 2016, there is no economical substitute—steelmaking is one of the largest industrial contributors of the CO_2 emissions in the world (see greenhouse gases).

The challenge set by the greenhouse gas emissions of the blast furnace is being addressed in an ongoing European Program called ULCOS (Ultra Low CO_2 Steelmaking). Several new process routes have been proposed and investigated in depth to cut specific emissions (CO_2 per ton of steel) by at least 50%. Some rely on the capture and further storage (CCS) of CO_2, while others choose to decarbonize iron and steel production, by turning to hydrogen, electricity, and biomass. In the nearer term, a technology that incorporates CCS into the blast furnace process itself and is called the Top-Gas Recycling Blast Furnace is under development, with a scale-up to a commercial size blast furnace under way. The technology should be fully demonstrated by the end of the 2010s, in line with the timeline set, for example, by the EU to cut emissions significantly. Broad deployment could take place from 2020 on.

Terms

1. affinity 亲和性,亲和力
2. batch process 批量生产
3. silica 二氧化硅
4. calcium oxide 氧化钙
5. contamination 污染
6. flux 熔;熔剂
7. porous 透气的
8. permeable 可渗透的
9. crush 压碎
10. oxide 氧化物
11. decomposition 分解
12. acidic impurity 酸性杂质
13. equilibrium 平衡
14. desulphurisation 脱硫
15. decarbonize 脱碳
16. biomass 生物量

Exercises

1. Briefly discuss how the blast furnace operates.
2. Some "pig iron" has been used to produce cast iron. What else can it do?

READING & CRITICAL THINKING

LECTURE FIVE
Steelmaking

Passage A An Overview of Steelmaking

Steelmaking is the process of producing steel from iron ore and/or scrap. In steelmaking, impurities such as nitrogen, silicon, phosphorus, sulfur and excess carbon (most important impurity) are removed from the sourced iron, and alloying elements such as manganese, nickel, chromium, carbon and vanadium are added to produce different grades of steel. Limiting dissolved gases such as nitrogen and oxygen and entrained impurities (termed "inclusions") in the steel is also important to ensure the quality of the products cast from the liquid steel.

Steelmaking has existed for millennia, but it was not commercialized on a massive scale until the late 19th century. An ancient process of steelmaking was the crucible process. In the 1850s and 1860s, the Bessemer process and the Siemens-Martin process turned steelmaking into a heavy industry. Today there are two major commercial processes for making steel, namely basic oxygen steelmaking, which has liquid pig-iron from the blast furnace and scrap steel as the main feed materials, and electric arc furnace (EAF) steelmaking, which uses scrap steel or direct reduced iron (DRI) as the main feed materials. Oxygen steelmaking is fuelled predominantly by the exothermic nature of the reactions inside the vessel; in contrast, in EAF steelmaking, electrical energy is used to melt the solid scrap and/or DRI materials. In recent times, EAF steelmaking technology has evolved closer to oxygen steelmaking as more chemical energy is introduced into the process.

Steelmaking has played a crucial role in the development of ancient, medieval, and modern technological societies. Early processes of steel making were made during the classical era in India, China, Iran, and Rome, but the process of ancient steelmaking was lost in the West after the fall of the Western Roman Empire in the 5th century CE.

Cast iron is a hard, brittle material that is difficult to work, whereas steel is malleable, relatively easily formed and a versatile material. For much of the human history, steel has only been made in small quantities. Since the invention of the Bessemer process in the 19th century and subsequent technological developments in the injection technology and process control, mass production of steel has become an integral part of the global economy and a key indicator of the modern technological development. The earliest means of producing steel was in a bloomery.

An important aspect of the Industrial Revolution was the development of large-scale methods of producing forgeable metal (bar iron or steel). The puddling furnace was initially a means of producing wrought iron but was later applied to steel production.

The real revolution in modern steelmaking only began at the end of the 1850s when the Bessemer process became the first successful method of steelmaking in high quantity followed by the open-hearth furnace.

Modern steelmaking processes can be divided into two categories: primary and secondary.

Primary steelmaking involves converting liquid iron from a blast furnace and steel scrap into steel via basic oxygen steelmaking, or melting scrap steel or direct reduced iron (DRI) in an electric arc furnace.

Secondary steelmaking involves refining of the crude steel before casting, and the various operations are normally carried out in ladles. In secondary metallurgy, alloying agents are added, dissolved gases in the steel are lowered, and inclusions are removed or altered chemically to ensure that high-quality steel is produced after casting.

Primary steelmaking

Basic oxygen steelmaking is a method of primary steelmaking in which carbon-rich molten pig iron is converted into steel. Blowing oxygen through molten pig iron lowers the carbon content of the alloy and changes it into steel. The process is known as basic due to the chemical nature of the refractories—calcium oxide and magnesium oxide—that line the vessel to withstand the high temperature and corrosive nature of the molten metal and slag in the vessel. The slag chemistry of the process is also controlled to ensure that impurities such as silicon and phosphorus are removed from the metal.

The process was developed in 1948 by Robert Durrer, using a refinement of the Bessemer converter where blowing of air is replaced with blowing oxygen. It reduced the capital cost of the plants and the time of smelting, and increased the labor productivity. Between 1920 and 2000, labour requirements in the industry decreased by a factor of 1,000, from more than 3 man-hours per tonne to just 0.003 man-hours. The vast majority of steel manufactured in the world is produced using the basic oxygen furnace; in 2011, it accounted for 70% of global steel output. Modern furnaces will take a charge of iron of up to 350 tons and convert it into steel in less than 40 minutes compared to 10 – 12 hours in an open-hearth furnace.

Electric arc furnace steelmaking is the manufacture of steel from scrap or direct reduced iron melted by electric arcs. In an electric arc furnace, a batch of steel ("heat") may be started by loading scrap or direct reduced iron into the furnace, sometimes with a "hot heel" (molten steel from a previous heat). Gas burners may be used to assist with the melt down of the scrap pile in the furnace. As in basic oxygen steelmaking, fluxes are also added to protect the lining of the vessel and help remove impurities. Electric arc furnace steelmaking typically uses furnaces of capacity around 100 tonnes that produce steel every 40 to 50 minutes for further processing.

Secondary steelmaking

Secondary steelmaking is most commonly performed in ladles. Some of the operations performed in ladles include deoxidation (or "killing"), vacuum degassing, alloy addition, inclusion removal, inclusion chemistry modification, de-sulphurisation, and homogenisation. It is now common to perform ladle metallurgical operations in gas-stirred ladles with electric arc heating in the lid of the furnace. Tight control of ladle metallurgy is associated with producing high grades of steel in which the tolerances in chemistry and consistency are narrow.

In HIsarna ironmaking process, iron ore is processed almost directly into liquid iron or hot metal. The process is based around a type of blast furnace called a cyclone converter furnace, which makes it possible to skip the process of manufacturing pig iron pellets that is necessary for the basic oxygen steelmaking process. Without the necessity of this preparatory step, the HIsarna process is more energy-efficient and has a lower carbon footprint than traditional steelmaking processes.

Terms

1. iron scrap 废铁
2. dissolved gas 溶解气体
3. cast 浇铸
4. crucible process 坩埚加工
5. Bessemer process 酸性转炉炼钢法
6. Siemens-Martin process 平炉炼钢法
7. direct reduced iron 直接还原铁
8. crucible steel 坩埚钢
9. puddling furnace 搅炼炉
10. primary steelmaking 初炼钢
11. secondary steelmaking 精炼钢
12. crude steel 粗钢
13. ladle 长柄勺
14. alloying agent 合金添加剂
16. vacuum degassing 真空除气
17. cyclone converter furnace 旋风转炉
18. electric arc furnace steelmaking 电弧炉炼钢法
19. basic oxygen steelmaking 碱性氧气炼钢法

Exercises

1. What happens in the process of steelmaking?
2. How was steelmaking commercialized step by step?
3. Please explain the similarities and differences of primary steelmaking process and secondary steelmaking process.

Passage B Crucible Steel

Crucible steel is steel made by melting pig iron (cast iron), iron, and sometimes steel, often along with sand, glass, ashes, and other fluxes, in a crucible. In ancient times steel and iron were impossible to melt using charcoal or coal fires, which could not produce temperatures high enough. However, pig iron, having a higher carbon content thus a lower melting point, could be melted, and by soaking wrought iron or steel in the liquid pig iron for a long time, the carbon content of the pig iron could be reduced as it slowly diffused into the iron. Crucible steel of this type was produced in South and Central Asia during the medieval era. This generally produced a very hard steel, but also a composite steel that was inhomogeneous, consisting of a very high-carbon steel (formerly the pig iron) and a lower-carbon steel (formerly the wrought iron). This often resulted in an intricate pattern when the steel was forged, filed, or polished, with possibly the most well-known examples coming from the wootz steel used in Damascus swords. The steel was often much higher in carbon content and in quality (lacking impurities) in comparison with other methods of steel production of the time because of the use of fluxes.

Techniques for production of high-quality steel were developed by Benjamin Huntsman in England in the 18th century. Huntsman used coke rather than coal or charcoal, achieving temperatures high enough to melt steel and dissolve iron. Huntsman's process differed from some of the wootz processes in that it took a longer time to melt the steel and to cool it down and allowed more time for the diffusion of carbon. Huntsman's process used iron and steel as raw materials, in the form of blister steel, rather than direct conversion from cast iron as in puddling or the later Bessemer process. The ability to fully melt the steel removed any inhomogeneities in the steel, allowing the carbon to dissolve evenly into the liquid steel and negating the prior need for extensive blacksmithing in an attempt to achieve the same result. Similarly, it allowed steel to simply be poured into molds, or cast, for the first time. The homogeneous crystal structure of this cast steel improved its strength and hardness in comparison with preceding forms of steel. The use of fluxes allowed nearly complete extraction of impurities from the liquid, which could then simply float to the top for removal. This produced the first steel of modern quality, providing a means of efficiently changing excess wrought iron into useful steel. Huntsman's process greatly increased the European output of quality steel suitable for use in items like knives, tools, and machinery, helping to pave the way for the Industrial Revolution.

Methods of crucible steel production

Iron alloys are most broadly divided by their carbon content: cast iron has 2% — 4% carbon impurities; wrought iron oxidises away most of its carbon, to less than 0.1%. The much more valuable steel has a delicately intermediate carbon fraction, and its material properties range according to the carbon percentage: high carbon steel is stronger but more brittle than low carbon steel. Crucible steel sequesters the raw input materials from the heat source, allowing precise control of carburization (raising) or oxidation (lowering carbon content). Fluxes, such as limestone, could be added to the crucible to remove or promote sulfur, silicon, and other impurities, further altering its material qualities.

Previous to Huntsman, the most common method of producing steel was the manufacture of shear steel. In this method, blister steel produced by cementation was used, which consisted of a core of wrought iron surrounded by a shell of very high-carbon steel, typically ranging from 1.5% to 2.0% carbon. To help homogenize the steel, it was pounded into flat plates. This produced steel with alternating layers of steel and iron. The resulting billet could then be hammered flat, cut into plates, which were stacked and welded again, thinning and compounding the layers, and evening out the carbon more as it slowly diffused out of the high-carbon steel into the lower-carbon iron. However, the more the steel was heated and worked, the more it tended to decarburize, and this outward diffusion occurs much faster than the inward diffusion between layers. Thus, further attempts to homogenize the steel resulted in a carbon content too low for use in items like springs, cutlery, swords, or tools. Therefore, steel intended for use in such items, especially tools, was still being made primarily by the slow and arduous bloomery process in very small amounts and at high cost, which, albeit better, had to be manually separated from the wrought iron and was still impossible to fully homogenize in the solid state.

Huntsman's process was the first to produce a fully homogeneous steel. Unlike previous methods of steel production, the Huntsman process was the first to fully melt the steel, allowing the full diffusion of carbon throughout the liquid. With the use of fluxes, it also allowed the removal of most impurities, producing the first steel of modern quality. Due to carbon's high melting point (nearly triple that of steel) and its tendency to oxidise (burn) at high temperatures, it cannot usually be added directly to molten steel. However, by adding wrought iron or pig iron, allowing it to dissolve into the liquid, the carbon content could be carefully regulated (in a way similar to Asian crucible steel but without the stark inhomogeneities indicative of those steels). Another benefit was that it allowed other elements to be alloyed with the steel. Huntsman was one of the first to begin experimenting with the addition of alloying agents like manganese to help remove impurities such as oxygen from the steel. His process was later used by many others, such as Robert Hadfield and Robert Forester Mushet, to produce the first alloy steels like mangalloy, high-speed steel, and stainless steel.

Due to variations in the carbon content of the blister steel, the carbon steel produced could vary in carbon content between crucibles by as much as 0.18%, but on average produced a eutectoid steel containing 0.79% carbon. Due to the quality and high hardenability of the steel, it was quickly adopted for the manufacture of tool steel, machine tools, cutlery, and many other items. Because no oxygen was blown through the steel, it exceeded Bessemer steel in both quality and hardenability, so Huntsman's process was used for manufacturing tool steel until better methods, utilizing an electric arc, were developed in the early 20th century.

19th- and 20th-century production

In another method, developed in the United States in the 1880s, iron and carbon were melted together directly to produce crucible steel. Throughout the 19th century and into the 1920s a large amount of crucible steel was directed into the production of cutting tools, where it was called tool steel.

The crucible process continued to be used for specialty steels, but in today obsolete. Similar quality steels are now made with an electric arc furnace. Some uses of tool steel were displaced, first by high-speed steel and later by materials such as tungsten carbide.

Crucible steel elsewhere

Another form of crucible steel was developed in 1837 by the Russian engineer, Pavel Anosov. His technique relied less on the heating and cooling, and more on the quenching process of rapidly cooling the molten steel when the right crystal structure had formed within. He called his steel bulat; its secret died with him. In the United States crucible steel was pioneered by William Metcalf.

Terms

1. coal fire 煤火
2. melting point 熔点
3. composite steel 复合钢
4. file 锉;把……锉光
5. polish 打磨
6. Wootz steel 乌兹钢
7. Damascus sword 大马士革剑
8. blister steel 渗碳钢
9. inhomogeneity 异质物
10. blacksmithing 锻造
11. mold 模具
12. extraction 提取
13. carburization 渗碳(作用)
14. shear steel 剪切钢,优质切削钢
15. pound 敲打
16. flat plate 平板;浅平盘
17. stack 堆叠
18. compound 混合
19. decarburize 脱碳
20. diffusion 扩散
21. molten steel 钢水
22. mangalloy 锰钢
23. high-speed steel 高速钢
24. eutectoid steel 共析钢
25. specialty steel 特种钢
26. tungsten carbide 碳化钨

Exercises

1. What is the crucible steel?
2. How did Huntsman's process of producing a fully homogeneous steel differ from previous methods of steel production?
3. Does the crucible process continue to be used today?

Passage C Bessemer Process

The Bessemer process was the first inexpensive industrial process for the mass production of steel from molten pig iron before the development of the open-hearth furnace. The key principle is the removal of impurities from the iron by oxidation with air being blown through the molten iron. The oxidation also raises the temperature of the iron mass and keeps it molten.

Related decarburizing with air processes had been used outside Europe for hundreds of years, but not on an industrial scale. One such process (similar to puddling) was known in the 11th century in East Asia, where the scholar Shen Kuo of that era described its use in the Chinese iron and steel industry. In the 17th century, accounts by European travelers detailed its possible use by the Japanese.

The modern process is named after its inventor, the Englishman Henry Bessemer, who took out a patent on the process in 1856. The process was said to be independently discovered in 1851 by the American inventor William Kelly though the claim is controversial.

The process using a basic refractory lining is known as the "basic Bessemer process" or Gilchrist-Thomas process after the English discoverers Percy Gilchrist and Sidney Gilchrist Thomas.

History

A system akin to the Bessemer process has existed since the 11th century in East Asia. Economic historian Robert Hartwell writes that the Chinese of the Song Dynasty innovated a "partial decarbonization" method of repeated forging of cast iron under a cold blast. Sinologist Joseph Needham and historian of metallurgy Theodore A. Wertime have described the method as a predecessor to the Bessemer process of making steel. This process was first described by the prolific scholar and polymath government official Shen Kuo (1031 - 1095) in 1075, when he visited Cizhou. Hartwell states that perhaps the earliest center where this was practiced was the great iron-production district along the Henan-Hebei border during the 11th century.

In the 15th century the finery process, another process which shares the air-blowing principle with the Bessemer process, was developed in Europe. In 1740 Benjamin Huntsman developed the crucible technique for steel manufacture, at his workshop in the district of Handsworth in Sheffield. This process had an enormous impact on the quantity and quality of steel production, but it was unrelated to the Bessemer-type process employing decarburization.

The Japanese may have made use of a Bessemer-type process, which was observed by European travelers in the 17th century. The adventurer Johan Albrecht de Mandelslo describes the process in a book published in English in 1669. He writes, "They have, among others, particular invention for the melting of iron, without the using of fire, casting it into a tun done about on the inside without about half a foot of earth, where they keep it with continual blowing, take it out by ladles full, to give it what form they please." According to the historian Donald Wagner, Mandelslo did not personally visit Japan, so his description of the process is likely derived from accounts of other Europeans who had traveled to Japan. Wagner believes that the Japanese process may have been similar to the Bessemer process, but cautions that alternative explanations are also plausible.

In the early 1850s, the American inventor William Kelly experimented with a method similar to the Bessemer process. Wagner writes that Kelly may have been inspired by techniques introduced by Chinese ironworkers hired by Kelly in 1854. When Bessemer's patent for the process was reported by Scientific American, Kelly responded by writing a letter to the magazine. In the letter, Kelly states that he had previously experimented with the process and claimed that Bessemer knew of Kelly's discovery. He wrote that "I have reason to believe my discovery was known in England three or four years ago, as a number of English puddlers visited this place to see my new process. Several of them have since returned to England and may have spoken of my invention there".

Sir Henry Bessemer described the origin of his invention in his autobiography written in 1890. During the outbreak of the Crimean War, many English industrialists and inventors became interested in military technology. According to Bessemer, his invention was inspired by a conversation with Napoleon III in 1854, pertaining to the steel required for better artillery. Bessemer claimed that it "was the spark which kindled one of the greatest revolutions that the present century had to record, for during my solitary ride in a cab that night from Vincennes to Paris, I made up my mind to try what I could do to improve the quality of iron in the manufacture of guns." At the time steel was used to make only small items like cutlery and tools, but was too expensive for cannons. Starting in January 1855, he began working on a way to produce steel in the massive quantities required for artillery and by October he filed his first patent related to the Bessemer process. He patented the method a year later in 1856.

Bessemer licensed the patent for his process to four ironmasters, for a total of £ 27,000, but the licensees failed to produce the quality of steel he had promised—it was "rotten hot and rotten cold", according to his friend, William Clay—and he later bought them back for £ 32,500. His plan had been to offer the licenses to one company in each of several geographic areas, at a royalty price per ton that included a lower rate on a proportion of their output to encourage production, but not so large a proportion that they might decide to reduce their selling prices. By this method he hoped to cause the new process to gain in standing and market share.

He realized that the technical problem was due to impurities in the iron and concluded that the solution lay in knowing when to turn off the flow of air in his process so that the impurities were burned off but just the right amount of carbon remained. However, despite spending tens of thousands of pounds on experiments, he could not find the answer. Certain grades of steel are sensitive to the 78% nitrogen which was part of the air blast passing through the steel.

Bessemer was sued by the patent purchasers who couldn't get it to work. In the end Bessemer set up his own steel company because he knew how to do it, even though he could not convey it to his patent users. Bessemer's company became one of the largest in the world and changed the face of steelmaking.

The solution was first discovered by English metallurgist Robert Forester Mushet, who had carried out thousands of experiments in the Forest of Dean. His method was to first burn off, as far as possible, all the impurities and carbon, then reintroduce carbon and manganese by adding an exact amount of spiegeleisen. This had the effect of improving the quality of the finished product, increasing its malleability—its ability to withstand rolling and forging at high temperatures and making it more suitable for a vast array of uses. Mushet's patent ultimately lapsed due to Mushet's inability to pay the patent fees and was acquired by Bessemer. Bessemer earned over 5 million dollars in royalties from the patents.

The first company to license the process was the Manchester firm of W & J

Galloway, and they did so before Bessemer announced it at Cheltenham in 1856. They are not included in his list of the four to whom he refunded the license fees. However, they subsequently rescinded their license in 1858 in return for the opportunity to invest in a partnership with Bessemer and others. This partnership began to manufacture steel in Sheffield from 1858, initially using imported charcoal pig iron from Sweden. This was the first commercial production.

Industrial revolution in the United States

Alexander Lyman Holley contributed significantly to the success of Bessemer steel in the United States. His *Treatise on Ordnance and Armor* is an important work on contemporary weapons manufacturing and steelmaking practices. In 1862, he visited Bessemer's Sheffield works, and became interested in licensing the process for use in the US. Upon returning to the US, Holley met with two iron producers from Troy, New York, John F. Winslow and John Augustus Griswold, who asked him to return to the United Kingdom and negotiate with the Bank of England on their behalf. Holley secured a license for Griswold and Winslow to use Bessemer's patented processes and returned to the United States in late 1863.

The trio began setting up a mill in Troy, New York in 1865. The factory contained a number of Holley's innovations that greatly improved productivity over Bessemer's factory in Sheffield, and the owners gave a successful public exhibition in 1867. The Troy factory attracted the attention of the Pennsylvania Railroad, which wanted to use the new process to manufacture steel rail. It funded Holley's second mill as part of its Pennsylvania Steel subsidiary. Between 1866 and 1877, the partners were able to license a total of 11 Bessemer steel mills.

One of the investors they attracted was Andrew Carnegie, who saw great promise in the new steel technology after a visit to Bessemer in 1872, and saw it as a useful adjunct to his existing businesses, the Keystone Bridge Company and the Union Iron Works. Holley built the new steel mill for Carnegie, and continued to improve and refine the process. The new mill, known as the Edgar Thomson Steel Works, opened in 1875, and started the growth of the United States as a major world steel producer. Using the Bessemer process, Carnegie Steel was able to reduce the costs of steel railroad rails from \$100 per ton to \$50 per ton between 1873 and 1875. The price of steel continued to fall until Carnegie was selling rails for \$18 per ton by the 1890s. Prior to the opening of Carnegie's Thomson works, steel output in the United States totaled around 157,000 tons per year. By 1910, American companies were producing 26 million tons of steel annually.

William Walker Scranton, manager and owner of the Lackawanna Iron & Coal

Company in Scranton, Pennsylvania, had also investigated the process in Europe. He built a mill in 1876 using the Bessemer process for steel rails and quadrupled his production.

Bessemer steel was primarily used in the United States for railroad rails. During the construction of the Brooklyn Bridge, a major dispute arose over whether crucible steel should be used instead of the cheaper Bessemer steel. In 1877, Abram Hewitt wrote a letter urging against the use of Bessemer steel in the construction of the Brooklyn Bridge. Bids had been submitted for both crucible steel and Bessemer steel; John A. Roebling's sons submitted the lowest bid for Bessemer steel, but at Hewitt's direction, the contract was awarded to J. Lloyd Haigh Co.

Technical details

Using the Bessemer process, it took between 10 and 20 minutes to convert three to five tons of iron into steel—it used to take at least a full day of heating, stirring and reheating to achieve this.

Oxidation

The blowing of air through the molten pig iron introduces oxygen into the melt which results in oxidation, removing impurities found in the pig iron, such as silicon, manganese, and carbon in the form of oxides. These oxides either escape as gas or form a solid slag. The refractory lining of the converter also plays a role in the conversion—clay linings are used when there is little phosphorus in the raw material—this is known as the acid Bessemer process. When the phosphorus content is high, dolomite, or sometimes magnesite, linings are used in the alkaline Bessemer limestone process. These are also known as Gilchrist-Thomas converters, after their inventors, Percy Gilchrist and Sidney Gilchrist Thomas. To produce steel with desired properties, additives such as spiegeleisen (a ferromanganese alloy), can be added to the molten steel once the impurities have been removed.

Managing the process

When the required steel had been formed, it was poured into ladles and then transferred into moulds while the lighter slag was left behind. The conversion process, called the "blow", was completed in approximately 20 minutes. During this period the progress of the oxidation of the impurities was judged by the appearance of the flame issuing from the mouth of the converter. The modern use of photoelectric methods of recording the characteristics of the flame greatly aided the blower in controlling final product quality. After the blow, the liquid metal was decarburized to the desired point and other alloying materials were added, depending on the desired product.

A Bessemer converter could treat a "heat" (batch of hot metal) of 5 tons to 30 tons at a time. They were usually operated in pairs, one being blown while another was being filled or tapped.

Predecessor processes

By the early 19th century the puddling process was widespread. Until technological advances made it possible to work at higher heats, slag impurities could not be removed entirely, but the reverberatory furnace made it possible to heat iron without placing it directly in the fire, offering some degree of protection from the impurity of the fuel source. Thus, with the advent of this technology, coal began to replace charcoal fuel. The Bessemer process allowed steel to be produced without fuel, using the impurities of the iron to create the necessary heat. This drastically reduced the costs of steel production, but raw materials with the required characteristics could be difficult to find.

High-quality steel was made by the reverse process of adding carbon to carbon-free wrought iron, usually imported from Sweden. The manufacturing process, called the cementation process, consisted of heating bars of wrought iron together with charcoal for periods of up to a week in a long stone box. This produced blister steel. The blister steel was put in a crucible with wrought iron and melted, producing crucible steel. Up to 3 tons of expensive coke was burnt for each ton of steel produced. Such steel when rolled into bars was sold at £ 50 to £ 60 (approximately £ 3,390 to £ 4,070 in 2008) a long ton. The most difficult and work-intensive part of the process, however, was the production of wrought iron done in finery forges in Sweden.

This process was refined in the 18th century with the introduction of Benjamin Huntsman's crucible steelmaking techniques, which added an additional three hours firing time and required additional large quantities of coke. In making crucible steel, the blister steel bars were broken into pieces and melted in small crucibles, each containing 20 kg or so. This produced higher-quality crucible steel but increased the cost. The Bessemer process reduced the time needed to make steel of this quality to about half an hour while requiring only the coke needed initially to melt the pig iron. The earliest Bessemer converters produced steel for £ 7 a long ton, although it initially sold for around £ 40 a ton.

"Basic" vs. acidic Bessemer process

Sidney Gilchrist Thomas, a Londoner with a Welsh father, was an industrial chemist who decided to tackle the problem of phosphorus in iron, which resulted in the production of low-grade steel. Believing that he had discovered a solution, he contacted his cousin, Percy Gilchrist, who was a chemist at the Blaenavon Ironworks. The manager at the time, Edward Martin, offered Sidney equipment for large-scale testing and helped him draw up a patent that was taken out in May 1878. Sidney Gilchrist Thomas's invention consisted of using dolomite or sometimes limestone linings for the Bessemer converter rather than clay, and it became known as the "basic" Bessemer rather than the "acid" Bessemer process. An additional advantage was that the process formed more slag in the converter, and this could be recovered and used very profitably as a phosphate fertilizer.

Importance

In 1898, *Scientific American* published an article called "Bessemer Steel and Its Effect on the World", explaining the significant economic effects of the increased supply in cheap steel. They noted that the expansion of railroads into previously sparsely inhabited regions of the country had led to settlement in those regions, and had made the trade of certain goods profitable, which had previously been too costly to transport.

The Bessemer process revolutionized steel manufacture by decreasing its cost, from \$ 40 per long ton to \$ 6 – 7 per long ton, along with greatly increasing the scale and speed of production of this vital raw material. The process also decreased the labor requirements for steelmaking. Before it was introduced, steel was far too expensive to make bridges or the framework for buildings and thus wrought iron had been used throughout the Industrial Revolution. After the introduction of the Bessemer process, steel and wrought iron became similarly priced, and some users, primarily railroads, turned to steel. Quality problems, such as brittleness caused by nitrogen in the blowing air, prevented Bessemer steel from being used for many structural applications. Open-hearth steel was suitable for structural applications.

Steel greatly improved the productivity of railroads. Steel rails lasted ten times longer than iron rails. Steel rails, which became heavier as prices fell, could carry heavier locomotives, which could pull longer trains. Steel rail cars were longer and were able to increase the freight to car weight from 1: 1 to 2: 1.

As early as 1895 in the UK it was being noted that the heyday of the Bessemer process was over and that the open-hearth method predominated. *The Iron and Coal Trades Review* said that it was "in a semi-moribund condition. Year after year, it has not only ceased to make progress, but it has absolutely declined". It has been suggested, both at that time and more recently, that the cause of this was the lack of trained personnel and investment in technology rather than anything intrinsic to the process itself. For example, one of the major causes of the decline of the giant ironmaking company Bolckow Vaughan of Middlesbrough was its failure to upgrade its technology. The basic process, the Thomas-Gilchrist process, remained in use longer, especially in Continental Europe, where iron ores were of high-phosphorus content, and the open-hearth process was not able to remove all phosphorus; almost all inexpensive construction steel in Germany was produced with this method in the 1950s and 1960s. It was eventually superseded by basic oxygen steelmaking.

Obsolescence

In the US, commercial steel production using this method stopped in 1968. It was replaced by processes such as the basic oxygen (Linz-Donawitz) process, which offered better control of final chemistry. The Bessemer process was so fast (10 – 20 minutes for a heat) that it allowed little time for chemical analysis or adjustment of the alloying elements in the steel. Bessemer converters did not remove phosphorus efficiently from the molten steel; as low-phosphorus ores became more expensive, conversion costs increased. The process permitted only limited amount of scrap steel to be charged, further increasing costs, especially when scrap was inexpensive. Use of electric arc furnace technology competed favorably with the Bessemer process resulting in its obsolescence.

Basic oxygen steelmaking is essentially an improved version of the Bessemer process (decarburization by blowing oxygen as gas into the heat rather than burning the excess carbon away by adding oxygen carrying substances into the heat). The advantages of pure oxygen blast over air blast were known to Henry Bessemer, but the 19th-century technology was not advanced enough to allow for the production of the large quantities of pure oxygen to make it economically feasible for use.

Terms

1. artillery 大炮
2. alkaline 碱性的，含碱的
3. low-phosphorus ore 低磷矿石
4. malleability 可锻性；加工性；可塑性；适应性
5. magnesite 菱镁矿
6. Bessemer 贝塞麦(1813—1898 年)，英国工程师和发明家

Exercises

1. What is the main principle of the Bessemer method?
2. What is the process of making high-quality steel?
3. What is the basic Bessemer process?

Passage D Siemens-Martin Process

Sir William Siemens

Sir William Siemens (1823 - 1883), christened Carl Wilhelm, an eminent inventor, engineer, and natural philosopher, was born at Lenthe in Hanover on the 4th April, 1823. After being educated in the polytechnic school of Magdeburg and the university of Gottingen, he visited England at the age of nineteen, in the hope of introducing a process in electro-plating invented by himself and his brother Werner. The invention was adopted by Messrs Elkington, and Siemens returned to Germany to enter the engineering works of Count Stolberg at Magdeburg as a pupil. In 1844 he was again in England with another invention, the "chronometric" or differential governor for steam-engines (see STEAM-ENGINE). Finding that British patent laws afforded the inventor a protection which was then wanting in Germany, he thenceforth made England his home; but it was not till 1859 that he formally became a naturalized British subject. After some years spent in active invention and experiment at mechanical works near Birmingham, he went into practice as an engineer in 1851. He laboured mainly in two distinct fields, the applications of heat and the applications of electricity, and was characterized in a very rare degree by a combination of scientific comprehension with practical instinct. In both fields he played a part which would have been great in either alone; and, in addition to this, he produced from time to time miscellaneous inventions and scientific papers sufficient in themselves to have established a reputation. His position was recognized by his election in 1862 to the Royal Society, and later to the presidency of the Institute of Mechanical Engineers, the Society of Telegraph Engineers, the Iron and Steel Institute, and the British Association; by honorary degrees from the universities of Oxford, Glasgow, Dublin, and Wurzburg; and by knighthood. He died in London on the 19th of November, 1883.

Charging the open-hearth furnace

This is a reverberatory furnace to which a regenerative system of heating is connected . The capacity of the average open-hearth furnace is about 60 tons of metal.

Supposing the furnace to be at a moderate heat, ready for the charge, the tapping hole, which leads from the lowest part of the basin through the far side of the furnace, is stopped by ramming into it a quantity of magnesite from the outside.

The charge consists of pig iron, limestone, usually iron oxide, and steel scrap if any of this is available. A quantity of limestone, determined by experience, is first thrown in through the charging door. The pig iron is then brought molten in a large ladle from the mixer or directly from the blast furnace, and is transferred from the ladle to the furnace hearth by means of a portable refractory-lined trough. Sometimes solid pigs may be thrown in at the charging door, but this is not the best practice, as there is a considerable saving of fuel and handling by using molten pig. A quantity of steel scrap is next thrown in if available, but scrap must not be used unless its composition is known to be suitable to the grade of steel to be made. Steel scrap of the proper composition is highly desirable, as its impurities have been greatly reduced in its manufacture. A small quantity of iron ore, low in sulphur and phosphorus, is added to the charge.

Operation of the open-hearth furnace

The purpose of this operation is to remove, so far as can be done by the process, the silicon, manganese, carbon, phosphorus, and sulphur in the charge. The removal of sulphur is difficult and uncertain, phosphorus is removed only in the basic process, and the remaining ingredients are usually reduced below the quantities desired in the finished steel and are reintroduced at the end of the process.

The doors are tightly closed after charging, and the heat is regulated to melt the whole charge gradually, requiring from 2 to 4 hours. The furnace temperature begins at once to rise, and the limestone ($CaCO_3$) begins to decompose, forming CaO and CO_2. The increase of heat soon causes the silicon, manganese, and carbon to oxidise. The oxygen for this purpose is supplied mainly from the iron ore in the charge, to a less extent from the CO_2 of the lime, and to a small extent from the air entering through the regenerators. As the charge becomes more and more fluid, the iron and scrap become mixed, distributing their impurities evenly throughout their combined mass. The lime (CaO) and iron oxide float to the surface of the molten iron and become fused, mixing with the slag which has begun to form from the oxidised silicon and manganese and from the earthy matter of the charge. The slag spreads out evenly over the bath of iron, protecting it from the oxidizing action of the flame.

Unlike puddling, no stirring or rabbling is done, though the bottom of the basin is raked over by a long iron bar inserted through the small openings in the charging doors to loosen any part of the charge which may have stuck to the hearth.

From the time the metal is thoroughly melted, samples are occasionally dipped from the bath by means of a small ladle with a long handle. These samples are cast in a small iron mould, and when cold, are taken from the mould and broken. The melter judges instantly by inspection of the fracture the amount of carbon and phosphorus contained, and regulates the process accordingly. Another practice, more reliable, is to take the sample, after it has been cast and cooled, to the laboratory nearby and determine the quantity of these elements by exact chemical methods, requiring 15 or 20 minutes.

As the silicon and manganese decrease in the metal, the oxidation of carbon increases, causing the charge to "boil" as in wrought-iron making, due to the formation and escape of CO.

The melter watches the progress of the operation through peep holes in the furnace doors, protecting his eyes with dark-colored glasses. His experience enables him to regulate the furnace temperature to suit requirements. If the carbon is burning too fast, it is necessary to "pig up" the charge by adding solid pig to increase the carbon and chill the bath. If phosphorus (the last element to be attacked) is going too fast, compared with the carbon, as shown by the sampling, the consumption of carbon may be hastened by "oreing down", that is by adding iron to supply oxygen to consume the carbon. It is essential that an excess of iron oxide should not be added, particularly toward the end of the process, as an undue amount of iron oxide cannot be carried by the slag.

Toward the last of the process when the heat is still intense, and the bath is comparatively quiet from the cessation of other chemical action, the phosphorus is removed by becoming oxidised and at once combining with lime to form a stable compound. This compound, phosphate of lime, enters the slag. The burning out of the carbon continues at varying rates throughout the entire operation, and the last of it is not burned out until after the removal of the phosphorus.

When the carbon has been burned out, the purified metal has a higher melting point than before, and would "come to nature" or collect in plastic masses as in wrought-iron making were it not for the high heat of the furnace to keep it thoroughly fluid.

In making high-carbon steel it is the practice to stop the process when the carbon has burned out to just below the percent desired in the steel, and the small quantity needed is introduced by recarburizing as in the Bessemer process.

The elimination of sulphur is very irregular. It is the safest to use iron for the charge which has a sulphur content below that allowable in the steel, but this is not always practicable. Some sulphur will unite with lime and enter the slag, if very fluid, a condition assisted by throwing into the furnace a quantity of fluor-spar. Manganese ore added to the charge causes the formation of manganese sulphide, which also enters the slag.

The melter judges by the appearances in the furnace and particularly by the sampling, when the heat is finished.

Tapping out

It requires from 6 to 9 hours to bring a charge to the condition for tapping out. In this condition the bath of slag-covered metal contains some iron oxide and more or less oxygen, carbon monoxide, or other gases absorbed during the process. These must be removed so far as can be done, and the metal must be recarburized to give it the quantity of carbon needed to make the grade of steel desired. In the basic process the materials used to accomplish these results cannot be placed in the furnace in presence of the basic slag as they will reduce the phosphorus from the slag and cause it to reenter the metal, therefore these materials are mixed with the metal after it is tapped from the furnace.

The best material for this use is ferromanganese, as used in the Bessemer process. Calculation and experience determine the amounts of carbon and manganese needed for each furnace charge and the quantity of "ferro" necessary to give these amounts is heated and thrown into the metal as it flows into the ladle.

In some cases, the metal charge may be re-carburized by throwing pig iron into the furnace. Still another method of re-carburizing is to throw into the ladle the necessary quantity of pure coke or coal ground fine and held in paper bags, or a better way of distributing this form of carbon is to allow it to run into the ladle, as the metal runs in, from a hopper suspended above the ladle. Considerable experience is needed in using powdered carbon, to introduce the correct quantity, as some of it burns before it can be absorbed by the steel.

View along the "casting pit" of a set of open-hearth furnaces. F, backs of furnaces above charging platform P; C, traveling crane; A, jib or swinging cranes; R, runway for one end of large crane; L, ladles; M, groups of ingot moulds.

The manganese in the re-carburizer decomposes iron oxide, takes up oxygen in the metal, forming MnO, which floats to the surface as slag. Its action assists mechanically in removing some of the other gases and some of the slag in the metal.

A long steel bar is used to dig out the magnesite in the tapping hole. The metal flows into the ladle and nearly fills it. The slag flows from the furnace after the metal has flowed out, filling the ladle completely. Much of the slag runs over the edge of the ladle into a pit below, where it cools and is later lifted out by large hooks attached to the crane.

Pouring the moulds

When all the slag has flowed from the furnace, the crane lifts the ladle and carries it while the steel is teemed into the moulds, small pieces of aluminum are thrown into each mould with the steel, reducing a part of any remaining iron oxide it may contain. The aluminum also assists further to remove gases, which would cause blowholes in the steel.

The ground space adjacent to the row of furnaces on the tapping side is called the "casting pit". A long narrow pit is usually dug for holding the tall moulds in order that the workmen may remain at the ground level when performing their work during the pouring.

Terms

1. chronometric 精密计时的
2. basin 水池;流域;盆地
3. iron oxide 氧化铁
4. fluor-spar 萤石;氟石
5. manganese sulphide 硫化锰
6. ferromanganese 锰铁
7. reverberatory 反射的
8. crane 吊车,起重机
9. jib 船首三角帆;(起重机的)悬臂
10. blowhole (隧道等的)通风孔;岩石孔穴;[材] 气泡

Exercises

1. What is the purpose of adding a small amount of low-sulfur and low-phosphorus iron ore?

2. Where does the oxygen that causes the oxidation of silicon, manganese, and carbon in the furnace come from?

3. What is the function of throwing small pieces of aluminum into the mold?

Passage E Basic Oxygen Steelmaking

Basic oxygen steelmaking (BOS, BOP, BOF, or OSM), also known as Linz-Donawitz-steelmaking or the oxygen converter process is a method of primary steelmaking in which carbon-rich molten pig iron is made into steel. Blowing oxygen through molten pig iron lowers the carbon content of the alloy and changes it into the low-carbon steel. The process is known as basic because fluxes of burnt lime or dolomite, which are chemical bases, are added to promote the removal of impurities and protect the lining of the converter.

The process was developed in 1948 by Swiss engineer Robert Durrer and commercialized in 1952 - 1953 by the Austrian steelmaking company VOEST and ÖAMG. The LD converter, named after the Austrian towns Linz and Donawitz (a district of Leoben) is a refined version of the Bessemer converter where blowing of air is replaced with blowing oxygen. It reduced capital cost of the plants, the time of smelting, and increased labor productivity. Between 1920 and 2000, labor requirements in the industry decreased by a factor of 1,000, from more than three man-hours per metric ton to just 0.003. The majority of steel manufactured in the world is produced using the basic oxygen furnace. In 2000, it accounted for 60% of the global steel output.

Modern furnaces will take a charge of iron of up to 400 tons and convert it into steel in less than 40 minutes, compared to 10 - 12 hours in an open-hearth furnace.

History

The basic oxygen process developed outside of traditional "big steel" environment. It was developed and refined by a single man, Swiss engineer Robert Durrer, and commercialized by two small steel companies in allied-occupied Austria, which had not yet recovered from the destruction of World War II.

In 1856, Henry Bessemer patented a steelmaking process involving oxygen blowing for decarbonizing molten iron (UK Patent No. 2207). For nearly 100 years commercial quantities of oxygen were not available or were too expensive, and the invention remained unused. During WWII German (Karl Valerian Schwarz), Belgian (John Miles) and Swiss (Durrer and Heinrich Heilbrugge) engineers proposed their versions of oxygen-blown steelmaking, but only Durrer and Heilbrugge brought it to mass-scale production.

In 1943, Durrer, formerly a professor at the Berlin Institute of Technology, returned to Switzerland and accepted a seat on the board of Roll AG, the country's largest steel mill. In 1947, he purchased the first small 2.5-ton experimental converter from the US, and on April 3, 1948 the new converter produced its first steel. The new process could conveniently process large amounts of scrap metal with only a small proportion of primary metal necessary. In the summer of 1948 Roll AG and two Austrian state-owned companies, VOEST and ÖAMG, agreed to commercialize the Durrer process.

By June 1949, VOEST developed an adaptation of Durrer's process, known as the LD (Linz-Donawitz) process. In December 1949, VOEST and ÖAMG committed to building their first 30-ton oxygen converters. They were put into operation in November 1952 (VOEST in Linz) and May 1953 (ÖAMG, Donawitz) and temporarily became the leading edge of the world's steelmaking, causing a surge in steel-related research. Thirty-four thousand businesspeople and engineers visited the VOEST converter by 1963. The LD process reduced processing time and capital costs per ton of steel, contributing to the competitive advantage of Austrian steel. VOEST eventually acquired the rights to market the new technology. Errors by the VOEST and the ÖAMG management in licensing their technology made control over its adoption in Japan impossible. By the end of the 1950s, the Austrians lost their competitive edge.

In the original LD process, oxygen was blown over the top of the molten iron through the water-cooled nozzle of a vertical lance. In the 1960s, steelmakers introduced bottom-blown converters and introduced inert gas blowing for stirring the molten metal and removing phosphorus impurities.

In the Soviet Union, some experimental production of steel using the process was done in 1934, but industrial use was hampered by lack of efficient technology to produce liquid oxygen. In 1939, the Russian physicist Pyotr Kapitsa perfected the design of the centrifugal turboexpander. The process was put to use in 1942 - 1944. Most turboexpanders in industrial use since then have been based on Kapitsa's design and centrifugal turboexpanders have taken over almost 100% of the industrial gas liquefaction, and in particular the production of liquid oxygen for steelmaking.

Big American steelmakers were late adopters of the new technology. The first oxygen converters in the US were launched at the end of 1954 by McLouth Steel in Trenton, Michigan, which accounted for less than 1% of the national steel market. US Steel and Bethlehem Steel introduced the oxygen process in 1964. By 1970, half of the world's and 80% of Japan's steel output was produced in oxygen converters.

In the last quarter of the 20th century, use of basic oxygen converters for steel production was gradually, partially replaced by the electric arc furnace using scrap steel and iron. In Japan the share of LD process decreased from 80% in 1970 to 70% in 2000; worldwide share of the basic oxygen process stabilized at 60%.

Process

Basic oxygen steelmaking is a primary steelmaking process for converting molten pig iron into steel by blowing oxygen through a lance over the molten pig iron inside the converter. Exothermic heat is generated by the oxidation reactions during blowing.

The basic oxygen steelmaking process is as follows:

Molten pig iron (sometimes referred to as "hot metal") from a blast furnace is poured into a large refractory-lined container called ladle.

The metal in the ladle is sent directly for basic oxygen steelmaking or to a pretreatment stage. High purity oxygen at a pressure of 700 - 1,000 kilopascals (100 - 150 psi) is introduced at supersonic speed onto the surface of the iron bath through a water-cooled lance, which is suspended in the vessel and kept a few feet above the bath. Pretreatment of the blast furnace hot metal is done externally to reduce sulphur, silicon, and phosphorus before charging the hot metal into the converter. In external desulphurising pretreatment, a lance is lowered into the molten iron in the ladle and several hundred kilograms of powdered magnesium are added and the sulphur impurities are reduced to magnesium sulphide in a violent exothermic reaction. The sulfide is then raked off. Similar pretreatments are possible for external desiliconisation and external dephosphorisation using mill scale (iron oxide) and lime as fluxes. The decision to pretreat depends on the quality of the hot metal and the required final quality of the steel.

Filling the furnace with the ingredients is called charging. The BOS process is autogenous, i. e. the required thermal energy is produced during the oxidation process. Maintaining the proper charge balance, the ratio of hot metal from melt to cold scrap is important. The BOS vessel can be tilted up to 360° and is tilted towards the deslagging side for charging scrap and hot metal. The BOS vessel is charged with steel or iron scrap (25% - 30%), if required. Molten iron from the ladle is added as required for the charge balance. A typical chemistry of hotmetal charged into the BOS vessel is: 4% C, 0.2% - 0.8% Si, 0.08% - 0.18% P, and 0.01% - 0.04% S, all of which can be oxidised by the supplied oxygen except sulphur (which requires reducing conditions).

The vessel is then set upright and a water-cooled, copper tipped lance with 3 - 7 nozzles is lowered into it, and high purity oxygen is delivered at supersonic speeds. The lance "blows" 99% pure oxygen over the hot metal, igniting the carbon dissolved in the steel, to form carbon monoxide and carbon dioxide, causing the temperature to rise to about 1,700 ℃. This melts the scrap, lowers the carbon content of the molten iron and helps remove unwanted chemical elements. It is this use of pure oxygen (instead of air) that improves upon the Bessemer process, as the nitrogen (an undesirable element) and other gases in air do not react with the charge, and decrease efficiency of furnace.

Fluxes (burnt lime or dolomite) are fed into the vessel to form slag, to maintain basicity above 3 and absorb impurities during the steelmaking process. During "blowing", churning of metal and fluxes in the vessel forms an emulsion, that facilitates the refining process. Near the end of the blowing cycle, which takes about 20 minutes, the temperature is measured, and samples are taken. A typical chemistry of the blown metal is 0.3% - 0.9% C, 0.05% - 0.1% Mn, 0.001% - 0.003% Si, 0.01% - 0.03% S, and 0.005% - 0.03% P.

The BOS vessel is tilted towards the slagging side and the steel is poured through a tap hole into a steel ladle with basic refractory lining. This process is called tapping the steel. The steel is further refined in the ladle furnace, by adding alloying materials to impart special properties required by the customer. Sometimes argon or nitrogen is bubbled into the ladle to make the alloys mix correctly.

After the steel is poured off from the BOS vessel, the slag is poured into the slag pots through the BOS vessel mouth and dumped.

Variants

Earlier converters, with a false bottom that can be detached and repaired, are still in use. Modern converters have a fixed bottom with plugs for argon purging. The Energy Optimization Furnace (EOF) is a BOF variant associated with a scrap preheater where the sensible heat in the off-gas is used for preheating scrap, located above the furnace roof.

The lance used for blowing has undergone changes. Slagless lances, with a long tapering copper tip, have been employed to avoid the jamming of the lance during the blowing. Post-combustion lance tips burn the CO generated during the blowing into CO_2 and provide the additional heat. For the slag-free tapping, darts, refractory balls, and slag detectors are employed. Modern converters are fully automated with auto blowing patterns and sophisticated control systems.

Terms

1. oxygen converter process 氧气转炉炼钢法
2. burnt lime 煅石灰;氧化钙
3. basic oxygen furnace 碱性氧气转炉
4. scrap metal 废金属
5. primary metal 初生金属
6. molten iron 铁水;熔融铁
7. water-cooled nozzle 水冷喷嘴
8. lance 喷枪
9. gas liquefaction 气体液化
10. exothermic heat 放热
11. oxidation reaction 氧化反应
12. pretreatment 预处理
13. kilopascal 千帕
14. supersonic speed 超声速
15. water-cooled lance 水冷喷枪
16. magnesium sulphide 硫化镁
17. exothermic reaction 放热反应
18. desiliconisation 脱硅
19. dephosphorisation 脱磷
20. mill scale (iron oxide) 氧化皮
21. charging 填料
22. autogenous 自发的
23. thermal energy 热能
24. deslag 除渣
25. basicity 碱度;碱性
26. emulsion 乳剂
27. argon 氩气
28. molten pig iron (hot metal) 铁水
29. scrap preheater 废钢预热器
30. refractory-lined container 耐火材料衬里的容器
31. Energy Optimization Furnace (EOF) 能源优化炉
32. Linz-Donawitz-steelmaking 林茨·多纳维茨炼钢
33. big steel 钢铁巨头,包括美国几家大的钢铁公司,有美国钢铁公司(US Steel)、共和钢铁公司(Republic Steel)等8家公司
34. centrifugal turboexpander 离心式涡轮膨胀机

Exercises

1. What is the basic oxygen steelmaking?
2. How was the process of the basic oxygen steelmaking developed along the history?
3. Please explain the process of the basic oxygen steelmaking.

Passage F Electric Arc Furnace

An electric arc furnace (EAF) is a furnace that heats charged material by means of an electric arc.

Industrial arc furnaces range in size from small units of approximately one-ton capacity (used in foundries for producing cast iron products) up to about 400-ton units used for secondary steelmaking. Arc furnaces used in research laboratories and by dentists may have a capacity of only a few dozen grams. Industrial electric arc furnace temperatures can reach 1,800 ℃ (3,272 ℉), while laboratory units can exceed 3,000 ℃ (5,432 ℉).

Arc furnaces differ from induction furnaces, in that the charged material is directly exposed to an electric arc and the current in the furnace terminals passes through the charged material.

In the 19th century, a number of men had employed an electric arc to melt iron. Sir Humphry Davy conducted an experimental demonstration in 1810; welding was investigated by Pepys in 1815; Pinchon attempted to create an electrothermic furnace in 1853; and in 1878 - 1879, Sir William Siemens took out patents for electric furnaces of the arc type.

The first successful and operational furnace was invented by James Burgess Readman in Edinburgh, Scotland in 1888 and patented in 1889. This was specifically for the creation of phosphorus.

Further electric arc furnaces were developed by Paul Héroult, of France, with a commercial plant established in the United States in 1907. The Sanderson brothers formed the Sanderson Brothers Steel Co. in Syracuse, New York, installing the first electric arc furnace in the US. This furnace is now on display at Station Square, Pittsburgh, Pennsylvania.

Initially "electric steel" was a specialty product for such uses as machine tools and spring steel. Arc furnaces were also used to prepare calcium carbide for use in carbide lamps. The Stassano electric furnace is an arc type furnace that usually rotates to mix the bath. The Girod furnace is similar to the Héroult furnace.

While EAFs were widely used in World War II for the production of alloy steels, it was only later that electric steelmaking began to expand. The low capital cost for a mini-mill—around US $140 - 200 per ton of annual installed capacity, compared with

US $1,000 per ton of annual installed capacity for an integrated steel mill—allowed mills to be quickly established in war-ravaged Europe and allowed them to successfully compete with the big United States steelmakers, such as Bethlehem Steel and US Steel, for low-cost, carbon steel "long products" (structural steel, rod and bar, wire, and fasteners) in the US market.

When Nucor—now one of the largest steel producers in the US—decided to enter the long products market in 1969, they chose to start up a mini-mill, with an EAF as its steelmaking furnace, soon followed by other manufacturers. Whilst Nucor expanded rapidly in the Eastern US, the companies that followed them into mini-mill operations concentrated on local markets for long products, where the use of an EAF allowed the plants to vary production according to the local demand. This pattern was also followed globally, with EAF steel production primarily used for long products, while integrated mills, using blast furnaces and basic oxygen furnaces, cornered the markets for "flat products"—sheet steel and heavier steel plate. In 1987, Nucor made the decision to expand into the flat products market, still using the EAF production method.

Construction

An electric arc furnace used for steelmaking consists of a refractory-lined vessel, usually water-cooled in larger sizes, covered with a retractable roof, and through which one or more graphite electrodes enter the furnace. The furnace is primarily split into three sections: the shell, which consists of the sidewalls and lower steel "bowl"; the hearth, which consists of the refractory that lines the lower bowl; the roof, which may be refractory-lined or water-cooled, and can be shaped as a section of a sphere, or as a frustum (conical section). The roof also supports the refractory delta in its centre, through which one or more graphite electrodes enter.

The hearth may be hemispherical in shape, or in an eccentric bottom tapping furnace, the hearth has the shape of a halved egg. In modern meltshops, the furnace is often raised off the ground floor, so that ladles and slag pots can easily be maneuvered under either the end of the furnace. Separated from the furnace structure is the electrode support and electrical system, and the tilting platform on which the furnace rests. Two configurations are possible: the electrode supports and the roof tilt with the furnace, or are fixed to the raised platform.

A typical alternating current furnace is powered by a three-phase electrical supply and therefore has three electrodes. Electrodes are round in section, and typically in segments with threaded couplings, so that as the electrodes wear, new segments can be added. The arc forms between the charged material and the electrode, the charge is heated both by current passing through the charge and by the radiant energy evolved by the arc. The electric arc temperature reaches around 3,000 ℃ (5,432 ℉), thus causing the lower sections of the electrodes to glow incandescently when in operation. The electrodes are automatically raised and lowered by a positioning system, which may use either electric winch hoists or hydraulic cylinders. The regulating system maintains approximately constant current and power input during the melting of the charge, even though scrap may move under the electrodes as it melts. The mast arms holding the electrodes can either carry heavy busbars (which may be hollow water-cooled copper pipes carrying current to the electrode clamps) or be "hot arms", where the whole arm carries the current, increasing the efficiency. Hot arms can be made from the copper-clad steel or aluminium. Large water-cooled cables connect the bus tubes or arms with the transformer located adjacent to the furnace. The transformer is installed in a vault and is water-cooled.

The furnace is built on a tilting platform so that the liquid steel can be poured into another vessel for transport. The operation of tilting the furnace to pour molten steel is called "tapping". Originally, all steelmaking furnaces had a tapping spout closed with refractory that washed out when the furnace was tilted, but often modern furnaces have an eccentric bottom tap-hole (EBT) to reduce inclusion of nitrogen and slag in the liquid steel. These furnaces have a taphole that passes vertically through the hearth and shell, and is set off-centre in the narrow "nose" of the egg-shaped hearth. It is filled with refractory sand, such as olivine, when it is closed off. Modern plants may have two shells with a single set of electrodes that can be transferred between the two; one shell preheats scrap while the other shell is utilised for meltdown. Other DC-based furnaces have a similar arrangement, but have electrodes for each shell and one set of electronics.

AC furnaces usually exhibit a pattern of hot and cold-spots around the hearth perimeter, with the cold-spots located between the electrodes. Modern furnaces mount oxygen-fuel burners in the sidewall and use them to provide chemical energy to the cold-spots, making the heating of the steel more uniform. Additional chemical energy is provided by injecting oxygen and carbon into the furnace; historically this was done through lances (hollow mild-steel tubes) in the slag door, now this is mainly done through wall-mounted injection units that combine the oxygen-fuel burners and the oxygen or carbon injection systems into one unit.

A mid-sized modern steelmaking furnace would have a transformer rated about 60,000,000 volt-amperes (60 MVA), with a secondary voltage between 400 and 900 volts and a secondary current in excess of 44,000 amperes. In a modern shop such a furnace would be expected to produce a quantity of 80 metric tonnes of liquid steel in approximately 50 minutes from charging with cold scrap to tapping the furnace. In comparison, basic oxygen furnaces can have a capacity of 150 – 300 tonnes per batch, or "heat", and can produce a heat in 30 – 40 minutes. Enormous variations exist in furnace design details and operation, depending on the end product and local conditions, as well as ongoing research to improve the furnace efficiency. The largest scrap-only furnace (in terms of tapping weight and transformer rating) is a DC furnace operated by Tokyo Steel in Japan, with a tap weight of 420 metric tonnes and fed by eight 32 MVA transformers for 256 MVA total power.

To produce a ton of steel in an electric arc furnace requires approximately 400 kilowatt-hours per short ton or about 440 kWh per metric tonne; the theoretical minimum amount of energy required to melt a tonne of scrap steel is 300 kW, h (melting point 1,520 ℃/2,768 ℉). Therefore, a 300-tonne, 300 MVA EAF will require approximately 132 MWh of energy to melt the steel, and a "power-on time" (the time that steel is being melted with an arc) of approximately 37 minutes. Electric arc steelmaking is only economical where there is plentiful, reliable electricity, with a well-developed electrical grid. In many locations, mills operate during off-peak hours when utilities have surplus power generating capacity, and the price of electricity is less.

Operation

Scrap metal is delivered to a scrap bay, located next to the melt shop. Scrap generally comes in two main grades: shred (whitegoods, cars and other objects made of similar light-gauge steel) and heavy melt (large slabs and beams), along with some direct reduced iron (DRI) or pig iron for chemical balance. Some furnaces melt almost 100% DRI.

The scrap is loaded into large buckets called baskets, with "clamshell" doors for a base. Care is taken to layer the scrap in the basket to ensure good furnace operation; heavy melt is placed on top of a light layer of protective shred, on top of which is placed more shred. These layers should be present in the furnace after charging. After loading, the basket may pass to a scrap preheater, which uses hot furnace off-gases to heat the scrap and recover energy, increasing the plant efficiency.

The scrap basket is then taken to the melt shop, the roof is swung off the furnace, and the furnace is charged with scrap from the basket. Charging is one of the more dangerous operations for the EAF operators. A lot of potential energy is released by the tonnes of falling metal; any liquid metal in the furnace is often displaced upwards and outwards by the solid scrap, and the grease and dust on the scrap is ignited if the furnace is hot, resulting in a fireball erupting. In some twin-shell furnaces, the scrap is charged into the second shell while the first is being melted down, and preheated with off-gases from the active shell. Other operations are continuous charging—preheating scrap on a conveyor belt, which then discharges the scrap into the furnace proper, or charging the scrap from a shaft set above the furnace, with off-gases directed through the shaft. Other furnaces can be charged with hot (molten) metal from other operations.

After charging, the roof is swung back over the furnace and meltdown commences. The electrodes are lowered onto the scrap, an arc is struck, and the electrodes are then set to bore into the layer of shred at the top of the furnace. Lower voltages are selected for this first part of the operation to protect the roof and walls from excessive heat and damage from the arcs. Once the electrodes have reached the heavy melt at the base of the furnace, and the arcs are shielded by the scrap, the voltage can be increased, and the electrodes raised slightly, lengthening the arcs and increasing power to the melt. This enables a molten pool to form more rapidly, reducing tap-to-tap times. Oxygen is blown into the scrap, combusting or cutting the steel, and extra chemical heat is provided by wall-mounted oxygen-fuel burners. Both processes accelerate scrap meltdown. Supersonic nozzles enable oxygen jets to penetrate foaming slag and reach the liquid bath.

An important part of steelmaking is the formation of slag, which floats on the surface of the molten steel. Slag usually consists of metal oxides and acts as a destination for oxidised impurities and helps reduce erosion of the refractory lining. For a furnace with basic refractories, which includes most carbon steel-producing furnaces, the usual slag formers are calcium oxide (CaO, in the form of burnt lime) and magnesium oxide (MgO, in the form of dolomite and magnesite). These slag formers are either charged with the scrap, or blown into the furnace during the meltdown. Another major component of EAF slag is iron oxide from steel combusting with the injected oxygen. Later in the heat, carbon (in the form of coke or coal) is injected into this slag layer, reacting with the iron oxide to form metallic iron and carbon monoxide gas, which then causes the slag to foam, allowing greater thermal efficiency, and better arc stability and electrical efficiency. The slag blanket also covers the arcs, preventing damage to the furnace roof and sidewalls from radiant heat.

Once the scrap has completely melted down, and a flat bath is reached, another bucket of scrap can be charged into the furnace and melted down, although EAF development is moving towards single-charge designs. After the second charge is completely melted, refining operations take place to check and correct the steel chemistry and superheat the melt above its freezing temperature in preparation for tapping. More slag formers are introduced, and more oxygen is blown into the bath, burning out impurities such as silicon, sulfur, phosphorus, aluminium, manganese, and calcium, and removing their oxides to the slag. Removal of carbon takes place after these elements have burnt out first, as they have a greater affinity for oxygen. Metals that have a poorer affinity for oxygen than iron, such as nickel and copper, cannot be removed through oxidation and must be controlled through scrap chemistry alone, such as introducing the direct reduced iron and pig iron mentioned earlier. A foaming slag is maintained throughout and often overflows the furnace to pour out of the slag door into the slag pit. Temperature sampling and chemical sampling take place via automatic lances. Oxygen and carbon can be automatically measured via special probes that dip into the steel, but for all other elements, a "chill" sample—a small, solidified sample of the steel—is analysed on an arc-emission spectrometer.

Once the temperature and chemistry are correct, the steel is tapped out into a preheated ladle through tilting the furnace. For plain-carbon steel furnaces, as soon as the slag is detected during tapping, the furnace is rapidly tilted back towards the deslagging side, minimising slag carryover into the ladle. For some special steel grades, including stainless steel, the slag is poured into the ladle as well, to be treated at the ladle furnace to recover valuable alloying elements. During tapping, some alloy additions are introduced into the metal stream, and more lime is added on the top of the ladle to build a new slag layer. Often, a few tonnes of liquid steel and slag is left in the furnace to form a "hot heel", which helps preheat the next charge of scrap and accelerate its meltdown. During and after tapping, the furnace is "turned around": the slag door is cleaned of solidified slag, the visible refractories are inspected and water-cooled components checked for leaks, and electrodes are inspected for damage or lengthened through the addition of new segments; the taphole is filled with sand at the completion of tapping. For a 90-tonne, medium-power furnace, the whole process will usually take about 60 - 70 minutes from the tapping of one heat to the tapping of the next (the tap-to-tap time).

The furnace is completely emptied of steel and slag on a regular basis so that an inspection of the refractories can be made and larger repairs made if necessary. As the refractories are often made from calcined carbonates, they are extremely susceptible to hydration from water, so any suspected leaks from water-cooled components are treated extremely seriously, beyond the immediate concern of potential steam explosions. Excessive refractory wear can lead to breakouts, where the liquid metal and slag penetrate the refractory and furnace shell and escape into the surrounding areas.

Advantages for steelmaking

The use of EAFs allows steel to be made from a 100% scrap metal feedstock. This greatly reduces the energy required to make steel when compared with primary steelmaking from ores.

Another benefit is flexibility: while blast furnaces cannot vary their production by much and can remain in operation for years at a time, EAFs can be rapidly started and stopped, allowing the steel mill to vary the production according to the demand.

Although steelmaking arc furnaces generally use scrap steel as their primary feedstock, if hot metal from a blast furnace or direct-reduced iron is available economically, these can also be used as furnace feed.

As EAFs require large amounts of electrical power, many companies schedule their operations to take advantage of off-peak electricity pricing.

A typical steelmaking arc furnace is the source of steel for a mini-mill, which may make bars or strip product. Mini-mills can be sited relatively near the markets for steel products, so the transport requirements are less than for an integrated mill, which would commonly be sited near a harbor for better access to the shipping.

Issues

Although the modern electric arc furnace is a highly efficient recycler of steel scrap, operation of an arc furnace shop can have adverse environmental effects. Much of the capital cost of a new installation will be devoted to systems intended to reduce these effects, which include:

Enclosures to reduce high sound levels
Dust collector for the furnace off-gas
Slag production
Cooling water demand
Heavy truck traffic for scrap, materials handling, and product
Environmental effects of electricity generation

Because of the very dynamic quality of the arc furnace load, power systems may require technical measures to maintain the quality of power for other customers; flicker and harmonic distortion are common power system side-effects of the arc furnace operation. For this reason, the power station should be located as close as possible to the EA furnaces.

Other electric arc furnaces

For steelmaking, direct current (DC) arc furnaces are used, with a single electrode in the roof, and the current return through a conductive bottom lining or conductive pins in the base. The advantage of DC is lower electrode consumption per ton of steel produced, since only one electrode is used, as well as less electrical harmonics and other similar problems. The size of DC arc furnaces is limited by the current carrying capacity of available electrodes and the maximum allowable voltage. Maintenance of the conductive furnace hearth is a bottleneck in extended the operation of a DC arc furnace.

In a steel plant, a ladle furnace (LF) is used to maintain the temperature of liquid steel during the processing after tapping from EAF or to change the alloy composition. The ladle is used for the first purpose when there is a delay later in the steelmaking process. The ladle furnace consists of a refractory roof, a heating system, and, when applicable, a provision for injecting argon gas into the bottom of the melt for stirring. Unlike a scrap melting furnace, a ladle furnace does not have a tilting or scrap-charging mechanism.

Electric arc furnaces are also used for the production of calcium carbide, ferroalloys, and other nonferrous alloys and for the production of phosphorus. Furnaces for these services are physically different from steelmaking furnaces and may operate on a continuous, rather than batch, basis. Continuous-process furnaces may also use paste-type. Such a furnace is known as a submerged arc furnace, because the electrode tips are buried in the slag/charge, and arcing occurs through the slag, between the matte and the electrode. A steelmaking arc furnace, by comparison, arcs in the open. The key is the electrical resistance, which generates the heat required: the resistance in a steelmaking furnace is the atmosphere, while in a submerged-arc furnace, the slag (or charge) supplies the resistance. The liquid metal formed in either furnace is too conductive to form an effective heat-generating resistance.

Amateurs have constructed a variety of arc furnaces, often based on electric arc welding kits contained by silica blocks or flower pots. Though crude, these simple furnaces can melt a wide range of materials, create calcium carbide.

Cooling methods

- **Non-pressurized cooling system**

Smaller arc furnaces may be adequately cooled by circulation of air over structural elements of the shell and roof, but larger installations require intensive forced cooling to maintain the structure within safe operating limits. The furnace shell and roof may be cooled either by water circulated through pipes which form a panel, or by water sprayed on the panel elements. Tubular panels may be replaced when they become cracked or reach their thermal stress life cycle. Spray cooling is the most economical and is the highest efficiency cooling method. A spray cooling piece of equipment can be relined almost endlessly; equipment that lasts 20 years is the norm. However, while a tubular leak is immediately noticed in an operating furnace due to the pressure loss alarms on the panels, at this time there exists no immediate way of detecting a very small volume spray cooling leak. These typically hide behind slag coverage and can hydrate the refractory in the hearth leading to a break out of molten metal or in the worst case a steam explosion.

A plasma arc furnace (PAF) uses plasma torches instead of graphite electrodes. Each of these torches has a casing with a nozzle and axial tubing for feeding a plasma-forming gas (either nitrogen or argon) and a burnable cylindrical graphite electrode within the tubing. Such furnaces can be called "PAM" (Plasma Arc Melt) furnaces; they are used extensively in the titanium-melting industry and similar specialty metal industries.

Vacuum arc remelting (VAR) is a secondary remelting process for vacuum refining and manufacturing of ingots with improved chemical and mechanical homogeneity.

In critical military and commercial aerospace applications, material engineers commonly specify VIM-VAR steels. VIM means Vacuum Induction Melted and VAR means Vacuum Arc Remelted. VIM-VAR steels become bearings for jet engines, rotor shafts for military helicopters, flap actuators for fighter jets, gears in jet or helicopter transmissions, mounts or fasteners for jet engines, jet tail hooks and other demanding applications.

Most grades of steel are melted once and are then cast or teemed into a solid form prior to extensive forging or rolling to a metallurgically-sound form. In contrast, VIM-VAR steels go through two more highly purifying melts under vacuum. After melting in an electric arc furnace and alloying in an argon oxygen decarburization vessel, steels destined for vacuum remelting are cast into ingot molds. The solidified ingots then head for a vacuum induction melting furnace. This vacuum remelting process rids the steel of inclusions and unwanted gases while optimizing the chemical composition. The VIM operation returns these solid ingots to the molten state in the contaminant-free void of a vacuum. This tightly controlled melt often requires up to 24 hours. Still enveloped by the vacuum, the hot metal flows from the VIM furnace crucible into giant electrode molds. A typical electrode is about 15 feet (4.572 m) tall and will be in various diameters. The electrodes solidify under vacuum.

For VIM-VAR steels, the surface of the cooled electrodes must be ground to remove surface irregularities and impurities before the next vacuum remelt. Then the ground electrode is placed in a VAR furnace. In a VAR furnace, the steel gradually melts drop-by-drop in the vacuum-sealed chamber. Vacuum arc remelting further removes lingering inclusions to provide superior steel cleanliness and remove gases like oxygen, nitrogen and hydrogen. Controlling the rate at which these droplets form and solidify ensures a consistency of chemistry and microstructure throughout the entire VIM-VAR ingot, making the steel more resistant to fracture or fatigue. This refinement process is essential to meet the performance characteristics of parts like a helicopter rotor shaft, a flap actuator on a military jet, or a bearing in a jet engine.

For some commercial or military applications, steel alloys may go through only one vacuum remelt, namely the VAR. For example, steels for solid rocket cases, landing gears, or torsion bars for fighting vehicles typically involve one vacuum remelt.

Vacuum arc remelting is also used in the production of titanium and other metals which are reactive or in which high purity is required.

Terms

1. cast iron product 铸铁产品
2. steelmaking 炼钢
3. induction furnace 感应炉
4. electrothermic furnace 电热炉
5. patent 专利权
6. rack 齿轮
7. pinion drive 齿轮传动
8. refractory brick 耐火砖
9. calcium carbide 电石
10. carbide lamp 电石灯
11. annual installed capacity 年装机容量
12. fastener 扣件
13. basic oxygen furnace 碱性氧气转炉
14. molten bath 熔池
15. spout 出水管
16. retractable roof 可移动车顶
17. graphite electrode 石墨电极
18. sidewall 侧壁
19. slag pot 盛渣桶
20. configuration 配置
21. threaded coupling 螺纹接头
22. radiant energy 辐射能
23. positioning system 自动位置调节系统
24. electric winch hoist 电动绞车提升机
25. hydraulic cylinder 液压缸
26. bus bar 汇流排,江流条
27. electrode clamp 电极夹
28. molten steel 钢水
29. taphole 出铁口,出渣口

30. refractory sand 耐火砂
31. olivine 橄榄石
32. cold-spot 冷点
33. oxygen-fuel burner 氧气燃料烧嘴
34. slag door 渣门
35. volt-ampere 伏安
36. current 电流
37. electrical grid 输电网络
38. transformer vault 变压器室
39. hard hat 安全帽
40. dust mask 防尘面具
41. light-gauge steel 轻量型钢
42. melt shop 熔炼车间
43. grease 润滑油
44. twin-shell furnace 双壳炉
45. combust 燃烧
46. supersonic nozzle 超声速喷嘴
47. foaming slag 泡沫渣
48. liquid bath 熔池
49. metal oxide 金属氧化物
50. thermal blanket 隔热层
51. slag former 造渣剂
52. magnesium oxide 氧化镁
53. metallic iron 精炼铁
54. slag blanket 焊渣
55. slag pit 渣坑
56. temperature sampling 温度采集
57. chemical sampling 化学取样
58. automatic lance 自动喷枪
59. transformer 变压器
60. arc-emission spectrometer 电弧发射光谱仪
61. deslagging side 除渣侧
62. calcined carbonate 煅烧碳酸盐
63. scrap metal feedstock 废金属原料
64. primary feedstock 初级原料
65. frustum 平截头体
66. eccentric bottom tapping furnace 偏心底出钢炉
67. surplus power generating capacity 剩余发电容量
68. extensive dust collection system 大面积除尘系统
69. direct reduced iron (DRI) 直接还原铁
70. furnace feed 炉料
71. integrated mill 综合工厂,大型工厂,联合工厂
72. dust collector 除尘器
73. cooling water 冷却水
74. power system 电力系统
75. side-effect 副作用
76. direct current (DC) 直流电
77. electrical harmonics 电力谐波
78. ladle furnace (LF) 钢包炉
79. tapping 出钢
80. argon gas 氩气
81. submerged arc furnace 矿热炉
82. electrical resistance 电阻
83. heat-generating resistance 发热电阻
84. electric arc welding kit 电弧焊包
85. silica block 硅块
86. flower pot 花盆
87. furnace shell and roof 炉壳和炉顶
88. spray cooling 喷射冷却
89. hydrate 水合物
90. steam explosion 蒸汽喷发
91. plasma torch 等离子体焰炬
92. graphite electrode 石墨电极
93. axial tubing 轴管
94. vacuum refining 真空精炼
95. jet engine 喷射发动机

96. rotor shaft 转轴
97. flap actuator 翼板执行器
98. fighter jet 战斗机
99. mount 底座
100. jet tail hook 飞机尾钩
101. teem 浇注
102. contaminant-free 无污染的
103. helicopter rotor shaft 直升机旋翼轴
104. bearing 轴承
105. steel alloy 合金钢
106. vacuum remelt 真空重熔
107. solid rocket case 固体火箭壳体
108. landing gear 着陆装置
109. torsion bar 扭杆
110. off-peak electricity pricing 非高峰电价
111. non-pressurized cooling system 常压的冷却系统
112. plasma arc furnace (PAF) 等离子电弧炉
113. vacuum arc remelting (VAR) 真空电弧重熔
114. vacuum induction melted (VIM)真空感应熔化
115. vacuum arc remelted (VAR)真空电弧冶炼
116. argon oxygen decarburization vessel 氩氧脱碳器

Exercises

1. What is the electric arc furnace (EAF)?

2. What are the advantages of the use of EAF for steelmaking?

3. Are there any issues in the use of EAF for steelmaking? If there are, what are they?

READING & CRITICAL THINKING

LECTURE SIX

Situations of Iron and Steel in the World

Passage A Steel Industry in Europe, America, Japan and China

Steel is the world's most popular construction material because of its unique combination of durability, workability, and cost. It's an iron alloy that contains 0.2% —2% carbon by weight.

According to the World Steel Association, some of the largest steel-producing countries are China, India, Japan, and the US. China accounts for roughly 50% of this production. The world's largest steel producers include ArcelorMittal, China Baowu Group, Nippon Steel Corporation, and HBIS Group.

Methods for manufacturing steel have evolved significantly since the industrial production began in the late 19th century. Modern methods, however, are still based on the same premise as the original Bessemer process, which uses oxygen to lower the carbon content in iron.

Today, steel production makes use of recycled materials as well as traditional raw materials, such as iron ore, coal, and limestone. Two processes, basic oxygen steelmaking (BOS) and electric arc furnaces (EAF), account for virtually all steel production.

Ironmaking, the first step in making steel, involves the raw inputs of iron ore, coke, and lime being melted in a blast furnace. The resulting molten iron—also referred to as hot metal—still contains 4% - 4.5% carbon and other impurities that make it brittle.

Primary steelmaking has two methods: BOS (Basic Oxygen Furnace) and the more modern EAF (electric arc furnace) methods. The BOS method adds recycled scrap steel to the molten iron in a converter. At high temperatures, oxygen is blown through the metal, which reduces the carbon content to between 0 - 1.5%.

The EAF method, however, feeds the recycled steel scrap through high-power electric arcs (with temperatures of up to 1,650 degrees Celsius) to melt the metal and convert it into high-quality steel.

Secondary steelmaking involves treating the molten steel produced from both BOS and EAF routes to adjust the steel composition. This is done by adding or removing certain elements and/or manipulating the temperature and production environment. Depending on the types of steel required, the following secondary steelmaking processes can be used:

- Stirring
- Ladle furnace
- Ladle injection
- Degassing
- CAS-OB (composition adjustment by sealed argon bubbling with oxygen blowing)

Continuous casting sees the molten steel cast into a cooled mold, causing a thin steel shell to solidify. The shell strand is withdrawn using guided rolls, and then it's fully cooled and solidified. Next, the strand is cut depending on applications—slabs for flat products (plate and strip), blooms for sections (beams), billets for long products (wires), or thin strips.

In primary forming, the steel that is cast is then formed into various shapes, often by hot rolling, a process that eliminates cast defects and achieves the required shape and surface quality. Hot rolled products are divided into flat products, long products, seamless tubes, and specialty products.

Finally, it's time for manufacturing, fabrication, and finishing. Secondary forming techniques give the steel its final shape and properties. These techniques include:

- Shaping (cold rolling), which is done below the metal's recrystallization point, meaning mechanical stress—not heat—affects change
- Machining (drilling)
- Joining (welding)
- Coating (galvanizing)
- Heat treatment (tempering)
- Surface treatment (carburizing)

Terms

1. ArcelorMittal 安塞乐米塔尔钢铁公司
2. HBIS Group 河钢集团
3. ladle furnace 钢包精炼炉
4. injection 喷射
5. degas 脱气
6. continuous casting 连铸
7. steel shell 钢壳
8. slab 板材
9. bloom 坯块
10. seamless tube 无缝钢管
11. cold rolling 冷轧
12. galvanize 给(铁或钢)镀锌
13. surface treatment 表面处理
14. composition adjustment by sealed argon bubbling (CAS) 密闭式吹氩成分微调法

Exercises

1. What makes steel the world's most popular construction material?

2. What does BOS refer to? How does the method reduce the carbon content of the metal?

3. What are the major differences between hot rolled steel and cold rolled steel?

Passage B　Steel Industry in Europe

Europe's steel industry accounts for roughly nine percent of the global crude steel production. Europe is home to some of the oldest and largest steel companies in the world such as ArcelorMittal, headquartered in Luxembourg City. The production of crude steel in the EU has remained stable despite shifting global trade and markets. Construction, automotive manufacturing, and mechanical engineering account for the main part of Europe's steel consumption.

The history of industrial steel production in Europe dates back as early as the 19th century in some European countries, such as Britain and Germany. Like many other regions, the crude steel production was closely linked to the production of coal. One of Europe's largest producers of steel, Italy, was a world pioneer in the development of electric furnaces in the 20th century.

In more recent times, cheaper steel production in other countries has undercut the European steel industry. Since Europe has maintained a high steel demand, the European Union is among the leading importers of steel products in the world. Simultaneously, the EU is the leading exporter of scrap steel. Increased competition led to the collapse of many European steel producers in the 1950s, and ThyssenKrupp had to sell its elevator unit in 2020. The oil crisis of 1973 and the 1980s-recession led to a further decline in the European steel industry. At the beginning of 2020, the COVID-19 crisis led to a decline in steel production worldwide. While China's production output rebounded in the second quarter, resulting in a net year-on-year increase as of June 2020, Europe's production overall declined by 20 percent between June 2019 and June 2020.

Germany

Germany, with annual production of almost 43 million tons of crude steel in 2014, is the world's seventh-largest steel producer and the largest in the European Union (EU 28). Germany is responsible for 2.6 percent of world production, or a quarter of crude steel production in the EU. With 17.2 billion euros, the steel industry in Germany is responsible for about 30 percent of the value creation achieved by the European steel industry.

About two-thirds of steel in Germany is produced in integrated steel mills (blast furnace, steelworks, and rolling mill), the remaining third via the electric steel route. The manufacture of hot-rolled finished products totaled 36.5 m. tones in 2014. Most were flat products (65 percent), with long products making up the remaining 35 percent. Stainless and alloyed steels make up over 50 percent of the total production and thus have a higher status here than is internationally usual (approx. 30 percent). North Rhine-Westphalia is the German state that produces the most steel—about 40 percent.

As a basic industry, the steel sector is particularly important for the value-creation chains in Germany. The numerous innovations implemented by this industry and its close interrelations with other industrial sectors contribute towards the success of, for example, the car industry or machine construction. The steel sector supplies about one-fifth of the input purchases for machine construction and 12 percent of those for the automotive industry. Other important customer sectors include electrical engineering, the building sector as well as steel and metal processing. With about 3.5 million employees, the steel-intensive sectors account for two out of every three jobs in the processing industry.

The steel industry is also an important customer for numerous supplier sectors. This is due to its high input intensities: the steel industry is responsible for an annual transport volume of about 145 m. tons. Then there are the long production chains and the comprehensive range in production, as well as the associated services involved—from pig iron production to the rolled steel product. Studies on economic significance show that every euro of additional value creation in the steel industry in Germany generates a further 1.7 euros of value creation in upstream sectors. It has also been empirically proved that each job in the steel industry is connected with five to six further jobs in the supplier industries.

France

According to the statistics released by the French Ministry of Economy, Finance and Industry, in the first nine months this year France's basic steel product and ferroalloy imports amounted to a value of €4.92 billion, decreasing by 27.9 percent year on year.

In the given period, France imported € 1.16 billion of steel pipes and tubes—down 18.3 percent, €225.42 million of cold rolled steel bars—down 33.4 percent, €430.41 million of cold rolled steel strip—falling by 20.9 percent, €266.98 million of cold drawn wire—declining by 17.8 percent, and 1.19 billion of metal structures and parts—decreasing by 15 percent, all compared to the same period in 2019.

The course of industrialization in France was so idiosyncratic that for a longtime people wondered whether an industrial revolution had ever taken place in the country. One of the main reasons for this was that the "Grande Nation" did not possess as large and accessible natural supplies of coal and iron ore as countries like Great Britain or Belgium. Coal, in particular, was always a scarce commodity; the result was that the French relied on timber for an astonishingly long time. In addition, French agriculture functioned extraordinarily well. The 1789 revolution freed farmers and peasants from debts and taxes, thereby guaranteeing them a comparatively secure existence. The result was a lack of superfluous workers, a fact which gave a particular boost to the Industrial Revolution in Great Britain.

That said, there was a large variety of highly developed trades in 18th-century France. This can often be attributed to the wishes and demands of the aristocracy in the "ancien régime". Furniture and porcelain, leather goods and silk were manufactured in great style; and for many years French clocks were reputed to be the most precise in the world. The first person to process cloth on sewing machines was also a Frenchman. But this proved highly unfortunate for Barthélemy Thimonnier, because angry tailors burnt down his factory in Paris in 1830.

Industrialization set in hesitantly, not least boosted by the measures introduced by the State after the 1789 revolution. The introduction of the "code civil" occurred simultaneously with the abolition of the old guild restrictions and internal customs tariffs. The currency was stabilized and the Bank of France created. The state was involved in the construction of roads and canals.

But France remained primarily an agricultural country until way into the 20th century. Large new factory areas were concentrated in specific regions, above all in the north and east of the country. By 1830 there were three established cotton mill centres: around Rouen in Normandy, between Lille and Roubaix in the North, and the most modern in Alsace. In Mühlhausen this led to a highly efficient engineering industry which went on to export spinning machines and cotton looms to the whole of Europe.

Around the mid-19th century intensive coalmining sprang up in the Pas-de-Calais region. The other major coalmining area was in Lorraine. For several generations, the industry there was dominated by the De Wendel family. They owned coal mines and iron mills, and were highly committed to the new techniques coming out of England. They introduced both the steam engine and puddling kilns at a relatively early point in time. The latter dramatically improved the quality of the iron.

But the most famous ironworks in France was in Le Creusot. In 1784, it was one of the largest state creations in France, alongside a representative salkworks and glass-manufacturing factory. But it only really became significant when an entrepreneur by the name of Eugène Schneider took it over in 1836. It was about then that the Industrial Revolution in France really began to take off, not least because of the beginning of the railways. It was not long before the first French locomotives were being built in Le Creusot, and the Schneider family began to lay the foundations for their empire with rails and weapons.

The immense influence of the ancien régime made itself felt with the introduction of the motorcar at the end of the century. This luxury article was above all purchased by members of the aristocracy and major banking families in Paris. French companies ensured the spread of motor manufacturing, for firms like Peugeot, Panhard et Levassor (and shortly afterwards Renault) manufactured and sold motorcars in much greater volume than the small workshops of inventors in Germany were able to do. The French also supplied wealthy buyers in the important British market and finally captured a large share of the market by producing cars for the growing middle-class market.

The UK

The industrialization of steel production originated in the UK and was part of a variety of new manufacturing processes that helmed the Industrial Revolution.

However, though the UK was a leader in steel fabrication in the early 19th century, in the years since, its role on the global market declined.

In 2019, nearly two billion metric tons of crude steel were produced worldwide. As of 2018, China accounted for over half of all steel produced that year, while the EU28 made up less than 10 percent of the market.

Since the financial crash in 2009, production in the UK has dropped to a low of 7.3 million tons in 2018, though 2013 and 2014 saw output increases. Also in 2018, the amount of pig iron manufactured totaled roughly 5.6 million metric tons, having decreased from a peak of 9.7 million tons produced in 2014. With production figures declining and more energy-efficient means introduced, the iron and steel industry consumed only 2,560 gigawatt hours of electricity that same year, compared to 6,348 gigawatt hours used up in 2000.

Fabrication of metals has seen similar developments over the past few years. The amount of wire rods produced came to 815 thousand metric tons in 2018, which was less than half of the total made in 2004. Production figures of hot rolled steel plates also echo this decline, with 433 thousand tons in output recorded in 2014. Since 2014, however, figures have shown a moderate increasing trend.

Regarding imports, nine million tons of iron ore were brought into the country in 2018, a decrease compared to the previous year. That same year, pig iron imports also decreased, amounting to 37,000 metric tons.

As for exports, scrap metal sales to other countries came to nearly 8.7 million tons in 2018, which was only slightly less than the peak of 8.8 million tons recorded in the previous year. The main importer of the UK made iron and steel products was Turkey, having bought 760.7 million British pounds' worth of these items in 2018.

Terms

1. electric furnace 电炉
2. ThyssenKrupp 蒂森克虏伯股份公司
3. ferroalloy 铁合金
4. steel pipe 钢管
5. cold drawn wire 冷拔钢丝
6. wire rod 线材
7. hot rolled steel plate 热轧钢板

Exercises

1. Why does the European steel industry continue to decline?

2. Summarize the development of the steel industry in Germany, France and the UK.

3. According to the text, what are the main parts of steel consumption in European countries?

Passage C Steel Industry in America

A brief history of the American steel industry

Today, the currently operating US steel industry includes approximately 100 steel supply and steel production facilities, employing 140, 000 people, directly or indirectly supporting the livelihood of almost 1 million Americans. AHSS (advanced high-strength steel) is the only material that reduces greenhouse gas emissions in all phases of an automobile's life: manufacturing, driving, and end-of-life. Being the most recycled material in the world, copper, paper, glass, and plastic combined, over 60 million tons of steel are recycled or exported for recycling each year in North America alone.

The rise of the American steel industry

Early colonists had 2 primary goals: shelter and food. They needed to build homes, plant crops, and hunt. To facilitate these tasks, iron tools were needed, such as hammers, knives, saws, axes, nails, hoes, bullets, and horseshoes. Iron products were in demand, but it wasn't until the 19th century, when technological advances drove down the cost and increased the quality of the product, that steel manufacturing became a dominant industry. "With the abundant iron ore deposits around Lake Superior, the rich coal veins of Pennsylvania, and the easy access to cheap water transportation routes on the Great Lakes, the Midwest became the center of American heavy industry", business and financial historian John Steele Gordon writes in his "Importance of Steel" exposition. "In the years after the Civil War, the American steel industry grew with astonishing speed as the nation's economy expanded to become the largest in the world. Between 1880 and the turn of the century, American steel production increased from 1. 25 million tons to more than 10 million tons. By 1910, America was producing more than 24 million tons, by far the greatest of any country."

Strong technological foundation was the primary driving force behind the tremendous growth in the steel industry. Steel supply was crucial for rapid expansion of cities and urban infrastructure, railroads, bridges, factories, buildings. And eventually, in the 20th century, steel was used to make household appliances and automobiles. It was at this time that the US steel industry began using the open-hearth furnace, then the basic oxygen steelmaking process.

Long after World War II, the American steel service industry continued to flourish and serve as the foundation of the national economy. In 1969, American steel production peaked when the country produced 141,262,000 tons. Since then, large steel mills have been replaced by smaller mini-mills and specialty mills, using iron and steel scrap as feedstock, rather than iron ore.

American steel service and industry today

Although we've entered the computer age, American steel remains a top competitor in the global marketplace. It is the world's largest steel importer, according to the American Iron and Steel Institute, labor productivity has seen a five-fold increase since the early 1980s, going from an average of 10. 1 man-hours per finished ton to an average of 1. 9 man-hours per finished ton of steel in 2015. In addition, the North American steel industry is committed to the highest safety and health standards. Since 2005, the US steel producers have achieved a 70 percent reduction in both the total OSHA recordable injury and illness and lost workday case rates.

Advanced high-strength steel

The evolution of advanced high-strength steel (AHSS) continues to grow in application, notably in the automotive industry. New advanced steel grades are the lightweight automotive material that best addresses society's need for reduced greenhouse gas emissions, without compromising safety, performance, or affordability. These new steel types are already being used to improve the performance of vehicles on the road and emerging grades will be increasingly employed. Each year, new car models are introduced using lighter-weight, higher-strength steel components that provide a cost-effective answer to the demand for increased safety and fuel economy. Studies show that AHSS steel grades are growing faster in new automotive applications than aluminum and plastic—steel's main competitors.

Exercises

1. Iron products have long been available in the US, but why did the steel industry not become the dominant industry until the 19th century?

2. What was the driving force behind the rapid growth of the US steel industrys?
3. What are the advantages of AHSS over traditional materials?

Passage D Steel Industry in Japan

General situation of the steel industry in Japan

Japan is far from being self-sufficient in the iron and steel manufacture as carried on under modern conditions. The annual production of pig iron amounts to only a few hundred thousand tons, and the major part of this iron is smelted from foreign ores. The steel output is appreciably greater, and is likewise largely manufactured out of the imported pig iron and scrap. While Japan exports more coal than it imports, the native coal is not generally suitable for coking purposes, and hence the country is dependent upon foreign sources for its supply of coke and coking coal. In certain raw materials used in the production of alloy steel, like chromite, Japan is moderately well endowed. In the great basic materials, however, which are necessary for a large tonnage output, its natural resources are limited.

The consumption of steel in Japan vastly exceeds its production, and hence much steel is imported. The greater part of this imported material comes in the forms of steel plates and steel bars, rods and structural shapes. These forms are used largely in the building of ships, although much of this and similar material is employed in other industrial lines.

The large importation of iron ore, coking coal, and pig iron shows at the present time the steel industry of Japan. The dependence of the country upon foreign steel also indicates that the handicap of inadequate supplies of native raw materials is sufficient to keep the cost of steel manufacture at a relatively high point.

The largest producer in the country is the Imperial Steel Works located on the island of Kyushi near the ports of Wakamatsu and Moji. These works are owned and operated by the Japanese government. In addition to this government enterprise are several independent concerns, most of which are small and devoted to the manufacture of foundry products, but a few of considerable size and engaged in producing tonnage steel.

The rapid production increases of emerging steel-producing countries are mainly responsible for the recent growth; the world steel production expanded by no less than 180 million tons in the last three years. Thanks to the worldwide increase in the demand for steel, Japanese steelmakers have been operating at full capacity.

The main reasons for the high production level of the Japanese steel industry are the high activity level of the manufacturing industries of the country such as the automobile, shipbuilding and machinery industries, and the continued high level of steel exports to China and other Asian countries.

By far the greater part of the Japanese output of steel, and probably of pig iron, is manufactured in Japan proper. In recent years, however, there has been some movement of Japanese steel plants and foundries to the continent. While the present industrial depression has called a halt to this tendency, it is not improbable that in the near future this movement will again proceed.

The rapid growth in steel demand all over the world, however, has tightened the supply-demand balance of the raw materials for the industry. As a result, steel producers of various countries have been actively pursuing strategic alliances to secure a stable supply of the raw materials; agreements concluded recently for long-term iron ore supply are examples of such alliances.

With respect to environmental measures, on the other hand, the Japanese steel industry set forth a voluntary environmental action plan in 1996 under the coordination of the Japan Iron & Steel Federation (JISF), and is actively promoting countermeasures against global warming and the recycling of wastes.

The Japanese steel industry has experienced both boom and recession over the past 30 years. It is now undergoing a thorough restructuring. Like the British steel industry, it has faced serious economic, technical and political changes.

Japan-China cooperation in steel technology

Reviewing the cooperation between the steel industries of Japan and China, it has been 20 years since the No. 1 Blast Furnace of Baoshan Iron & Steel Co., Ltd. was first blown in Nippon Steel Corporation, assisting the construction of the works as a symbolic project of the Japan-China economic cooperation after the conclusion of the Treaty of Peace and Friendship between Japan and the People's Republic of China in 1978. Ever since, the steel industries of the two countries have enjoyed a continued friendly relationship. Moreover, the meetings of the Japan-China Steel Science Council have been held for more than 20 years under the joint auspices of the Iron & Steel Institute of Japan (ISIJ) and the Chinese Society for Metals (CSM).

The Japanese steel industry is also devising long-term strategies; the strategic road map of steel technology worked out by the government and the ISIJ is one such strategy.

Terms

1. chromite 铬铁矿；亚铬酸盐
2. steel bar 钢筋
3. ISIJ 日本钢铁协会
4. CSM 中国金属学会
5. JISF 日本钢铁联盟

Exercises

1. According to the passage, what do you think is hindering the growth of the Japanese steel industry?

2. Why is the movement of Japanese steel plants inevitable?

3. Why does the consumption of steel in Japan exceed its production? What are the consequences?

Passage E Steel Industry in China

The steel industry was small and sparsely populated at the start of the 20th century and during both world wars. Most of the steel infrastructure was destroyed during the wars, and was using Soviet technologies. China lagged behind the Western countries in its steel industry development even though they were using central planning techniques during the early days of communist rule.

China underwent rapid economic industrialization since Deng Xiaoping's reform and opening-up policy which took place four decades ago in 1978. The steel industry gradually increased its output. China's annual crude steel output was 100 million tons in 1996.

21st century

China produced 123 million tones (121,000,000 long tons; 136,000,000 short tons) of steel in 1999. After its ascension to the WTO, it aggressively expanded its production for its growing appetite of manufacturing industries such as automotive vehicles, consumer electronics, and building materials.

The Chinese steel industry is dominated by a number of large state-owned groups which are owned via shareholdings by local authorities, provincial governments, and even the central authorities. According to China Iron and Steel Association, the top 5 steel groups by production volume in 2015 are Baosteel Group—Wuhan Iron and Steel Corporation, Hesteel Group, Shagang Group, Ansteel Group, and Shougang Group.

By 2008 raw materials such as iron ore prices grew and China had to reluctantly agree to price increases by the three largest iron ore producers in the world: BHP Billiton, Rio Tinto, and Vale. During the global financial crisis, the Chinese steel mills won price reprieves as demand from their customers slowed. When the demand started to pick up again in 2009 and in 2010, the price crept back up due to the higher demand for automobiles, low interest rates, and government fiscal stimuli around the world. Prices for iron ore were negotiated on an annual contract pricing scheme. Australian iron ore producers were not happy that iron prices did not reflect spot market pricing. In 2010 pressure from BHP Billiton and Rio Tinto to move to a quarterly based index pricing succeeded. Many Japanese steel mills and Chinese steel companies had to follow as the demand for raw materials heated up. Spot-basis pricing has caused problems for steel manufacturers such as exposing them to price fluctuation in the market and reducing the stability of resource supply. Steel mills prefer long-term pricing to hedge against cost and maintain raw materials supply stability. Rio Tinto has said it will cancel contracts and sell the steel on the spot markets if Chinese steel mills back down on the new quarterly pricing regime.

In 2011 China was the largest producer of steel in the world, producing 45% of the world's steel, 683 million tons, an increase of 9% from 2010. 6 of 10 of the largest steel producers in the world are in China. Profits are low despite continued high demand due to high debt and overproduction of high-end products produced with the equipment financed by the high debt. The central government is aware of this problem but there is no easy way to resolve it as local governments strongly support the local steel production. Meanwhile, each firm aggressively increases production. Iron ore production kept pace with the steel production in the early 1990s but was soon outpaced by imported iron ore and other metals in the early 2000s. Steel production, an estimated 140 million tons in 2000 increased to 419 million tons in 2006. Much of the country's steel output came from a large number of small-scale producing centers, one of the largest being Anshan in Liaoning.

China was the top exporter of steel in the world in 2008. Export volumes in 2008 were 59.23 million tons, a 5.5% fall over the previous year. The decline ended China's decade-old steel export growth. As of 2012 steel exports faced widespread antidumping taxes and had not returned to pre-2008 levels. Domestic demand remained strong, particularly in the developing west where steel production in

Xinjiang was expanding.

On 26 April, 2012, a warning was issued by China's bank regulator to use caution with respect to lending money to steel companies who, as profits from the manufacture and sale of steel have fallen, have sometimes used borrowed money for speculative purposes. According to the China Iron and Steel Association, the Chinese steel industry lost 1 billion RMB in the first quarter of 2012, its first loss since 2000.

As of 2015 the global steel market was weak with both Ukraine and Russia attempting to export large amounts of steel. Weak domestic demand in 2014 resulted in record exports of 100 million metric tons of steel by the Chinese steel industry.

Efforts by the Chinese Ministry of Environmental Protection under the Action Plan for the Prevention and Control of Air Pollution has resulted in pressure on steel mills in Linyi and Chengde to employ environmental protection measures on pain of being closed down.

In the context of lowered demand (see also 2015—2016 Chinese stock market crash), in 2016 the Chinese state announced large-scale closures and redundancies in heavy and primary industries, many of which were functioning as zombie companies.

Exercises

1. Summarize the main features of the development of the early Chinese steel industry.

2. Considering the current state of the steel industry, what kind of technology do you think will be needed in the future? Why?

3. The steel industry has experienced both boom and recession over the past 30 years. Some people argue that traditional industries have declined. Find out the reasons and figure out how to solve them.